QUATERNARY FAULTS, FOLDS, AND SELECTED VOLCANIC FEATURES IN THE CEDAR CITY 1° X 2° QUADRANGLE, UTAH

By
R. Ernest Anderson
U.S. Geological Survey
and
Gary E. Christenson
Utah Geological and Mineral Survey

A primary mission of the UGMS is to provide geologic information of Utah through publications; the formal publication series is reserved for material whose senior author is a UGMS staff member. The Miscellaneous Publication series provides an outlet for non-UGMS authors without necessarily going through extensive policy, technical, and editorial review required by the formal series. It also provides a means for UGMS and non-UGMS authors to publish more interpretive work with the knowledge that readers will exercise some degree of caution.

UTAH GEOLOGICAL AND MINERAL SURVEY
a division of
UTAH DEPARTMENT OF NATURAL RESOURCES
MISCELLANEOUS PUBLICATION 89-6   **1989**

CONTENTS

FIGURES

QUATERNARY FAULTS, FOLDS, AND SELECTED VOLCANIC FEATURES IN THE CEDAR CITY 1° X 2° QUADRANGLE, UTAH

By
R. Ernest Anderson
U.S. Geological Survey
and
Gary E. Christenson
Utah Geological and Mineral Survey

ABSTRACT

The Cedar City 1°X 2° quadrangle in southwestern Utah is traversed by several major Quaternary structures and numerous lesser faults and folds which all trend generally north-northeast. The major structures include, from west to east, the Gunlock, Washington, Antelope Range, and Hurricane faults; the Cedar City-Parowan monocline; and the Red Hills, Paragonah, and Sevier faults. These structures generally range in age of last movement from middle to late Pleistocene. The most recent movement and highest slip rates occur on the Hurricane and Paragonah faults. Both are principally range-front faults which locally cut Quaternary volcanic rocks. Scarps on alluvium mappable at the 1:250,000 scale occur only on the Hurricane and Antelope Range faults. The Sevier fault is the longest of the major structures and is also principally a range-front fault.

Shorter, less prominent faults and folds are scattered throughout the quadrangle, but are concentrated along the northern edge in the Escalante Desert, Cedar and Parowan Valleys, and Sevier Valley. These structures are principally in alluvium and Quaternary basalt and are generally of middle to late Pleistocene age. The faults and folds are particularly well developed near Cedar City and in the Sevier Valley near Panguitch on the hanging wall blocks of the Hurricane and Sevier faults. The youngest (latest Pleistocene, possible Holocene) scarps on alluvium are near Enoch and in Parowan Valley. Other suspected young (late Pleistocene) scarps are in the Escalante Valley and in Sevier Valley north of Panguitch.

Few data are available to determine average recurrence intervals between surface-faulting events and the time of the last event. Long-term slip rates calculated from offset volcanic rocks of known age average about 5 m/m.y. for the Gunlock fault, 300-470 m/m.y. for the Hurricane fault, 455 m/m.y. for the Paragonah and associated faults, and 357 m/m.y. for the Sevier fault. Despite these relatively high long-term slip rates, the lack of evidence for Holocene faulting indicates that the probability of a surface-faulting earthquake at a particular point along any fault in the Cedar City quadrangle is relatively low compared to the Wasatch fault or other Holocene faults in northern Utah.

INTRODUCTION

The purpose of this map and report is to present selected aspects of the Quaternary structural history important for evaluating the potential for large surface-faulting earthquakes in the area covered by the Cedar City 1° X 2° quadrangle. Principal data for such an evaluation come from the study of known Quaternary faults and folds. Plate 1 shows the locations of all such structures. Fault categories are defined on the basis of the estimated age of most recent movement and type of material in which the fault is found. Fault categories include those with Holocene, late Pleistocene, middle to late Pleistocene, early to middle Pleistocene, or suspected Quaternary surface displacements. The type of material in which the fault is found and on which a scarp is formed includes unconsolidated deposits, chiefly alluvium; Quaternary or suspected Quaternary basalt; and pre-Quaternary bedrock either within range blocks or along linear, tectonically active range fronts and major block boundaries. Fault scarps on alluvium generally provide the most reliable indication of relative age of Quaternary movement. The latter categories are included to convey some information about young faulting in the nonalluviated highlands, especially in the high plateaus in the eastern part of the quadrangle. These areas are undergoing active erosion and the longevity of scarps in unconsolidated deposits is limited.

The erosional degradation of fault scarps on unconsolidated deposits and the techniques for measurement and analysis to estimate age of last movement are described in Wallace (1977), Anderson and Bucknam (1979a,b), and Bucknam and Anderson (1979). Relative ages can be assigned to single-event scarps by comparing their height versus maximum slope angle data to those from scarps of known or reliably inferred age (Bucknam and Anderson, 1979; Machette and others, 1986), by diffusion equation modeling (Nash, 1980; Hanks and others, 1984), or by their temporal relation to some regional datum such as the shorelines of Lake Bonneville. No regional time-stratigraphic datum exists in the Cedar City quadrangle, and the applicability of diffusion-equation modeling to scarps significantly older than Holocene, such as most scarps in the quadrangle, has not been established. No scarps young enough to compute a regression equation are present, and many scarps are so erosionally modified and old that profile data do not provide a reliable basis for estimating the age of last faulting. Therefore, relative ages of scarps on alluvium are determined by simply plotting scarp height versus maximum slope angle data. These data are then compared to regression lines for the 10,000-year-old Drum Mountains fault scarps and 15,000-year-old Bonneville shoreline scarps (Bucknam and Anderson, 1979). It is assumed that significant variations in climate and paleoclimate do not exist between areas of faulted alluvium in the Cedar City quadrangle and areas to the north where baseline scarp morphology data were gathered which, if present, would influence the rate of scarp degradation.

Estimating the age of last displacement on faults in Quaternary basalt and pre-Quaternary bedrock and along tectonically active mountain fronts lacking scarps is more difficult. Although evidence for faulting in basalt is generally well preserved, little can be inferred except that the faulting postdates eruption. Radiometric ages on faulted basalts are used to determine the maximum age of faulting. To constrain the time of the last movement, we assume that offsets of more than a few meters did not occur at one time but were spread in increments uniformly over the period of time following eruption.

Similar problems arise in assessing the timing of last movement on faults in pre-Quaternary bedrock. Criteria used in identifying Quaternary scarps on pre-Quaternary bedrock include: (1) qualitative photogeologic assessment of scarp linearity, steepness, and degree of erosional dissection by transverse drainages; (2) conspicuous oversteepening of stream gradients or anomalous erosional incision by transverse drainages where they cross scarps; and (3) steep or well-preserved scarps in relatively weak bedrock materials. Tectonically active range fronts are indicated by: (1) linearity, (2) faceted spurs, and (3) active alluvial-fan deposition valleyward of the range front. Age assignments for faults in basalt and bedrock and active range fronts are less certain than those in alluvium because: (1) basalt ages are not known everywhere, (2) elevation-related climatic contrasts will degrade scarps and range fronts at different rates in different areas, and (3) degradation rates of bedrock scarps will be different from those in alluvium and will differ among rocks of differing mechanical and chemical properties. No calibration data pertaining to these differences are available. Also, there is some uncertainty in discriminating by photogeologic methods between fault scarps and large linear landslide margins which commonly coincide in highland terrains. The record of Quaternary scarps on basalt and bedrock is probably more complete than the record on alluvium because alluvial scarps are more readily obliterated by erosion or burial. Conversely, scarps on alluvium are more recognizable on air photos and are more diagnostic of Quaternary movement. We believe that the introduction of these uncertainties is outweighed by the more comprehensive picture revealed by extending the portrayal of Quaternary faulting into bedrock areas.

Quaternary folds are common in the Cedar City quadrangle area. Because they are spatially, and probably genetically, related to Quaternary faults they are identified on the map and described in the text. Some are large-amplitude range-front structures such as the Cedar City-Parowan monocline, whereas others are much smaller features recognizable by associated faults (axial graben) or anomalously sloping alluvial surfaces. Still others involve originally flat-lying Quaternary basalt flows. All these folds are interpreted to be active during Quaternary time either because Quaternary-age deposits are folded or Quaternary-age faults are genetically associated with the folds.

Information for the map and report was gathered from a review of literature, interpretation of 1:60,000-scale air photos, and field reconnaissance in selected areas where faults with possible Quaternary displacement were identified. In addition, more detailed study locally included mapping of scarps on 1:20,000-scale air photos, topographic scarp profiling, and logging of exposures. Emphasis was placed on identifying principal seismogenic structures, however, and not on identifying all possible minor scarps and photolineaments in the quadrangle. The map thus represents a reconnaissance-level compilation of the well-expressed Quaternary faults and structures in the quadrangle. Investigations of some of the Quaternary features lead us to conclude that they are non-hazardous from a seismogenic standpoint, and these are described in a separate section.

No coseismic historical surface-fault ruptures are known in the quadrangle, and convincing evidence for Holocene coseismic faulting is absent. Most major faults in the quadrangle show evidence for late to latest Pleistocene movement, although the age of last movement generally cannot be determined with confidence. Reliable evidence for recurrence and the amount of displacement per event is generally lacking, and it can only be said that recurrence intervals for surface faulting along most faults appears to be greater than 10,000 years. This indicates that the probability of a large surface-faulting earthquake (Richter magnitude 6.5+) on the identified faults in the Cedar City quadrangle is low relative to areas such as the Wasatch Front.

PHYSIOGRAPHIC AND TECTONIC SETTING

The southern half of the Cedar City quadrangle is drained mainly by Beaver Dam Wash, the Virgin River, and Kanab Creek, all of which are part of the Colorado River system. This is an area of active downcutting and widespread landscape degradation. The northeastern part is drained mainly by streams of the Sevier River system which, although part of the interior drainage system of the Great Basin, form an integrated network that is actively downcutting. The northwestern part of the quadrangle is similar to other parts of the Great Basin and includes broad areas of Quaternary alluvium such as the Escalante Desert. Drainage is poorly developed, and overland flow occurs on the floors of the closed basins in much of the area.

The Cedar City quadrangle is traversed by the transitional boundary between the Colorado Plateau and Basin and Range structural provinces (figure 1). The southeasternmost part is in the stable Colorado Plateau interior. The northwestern

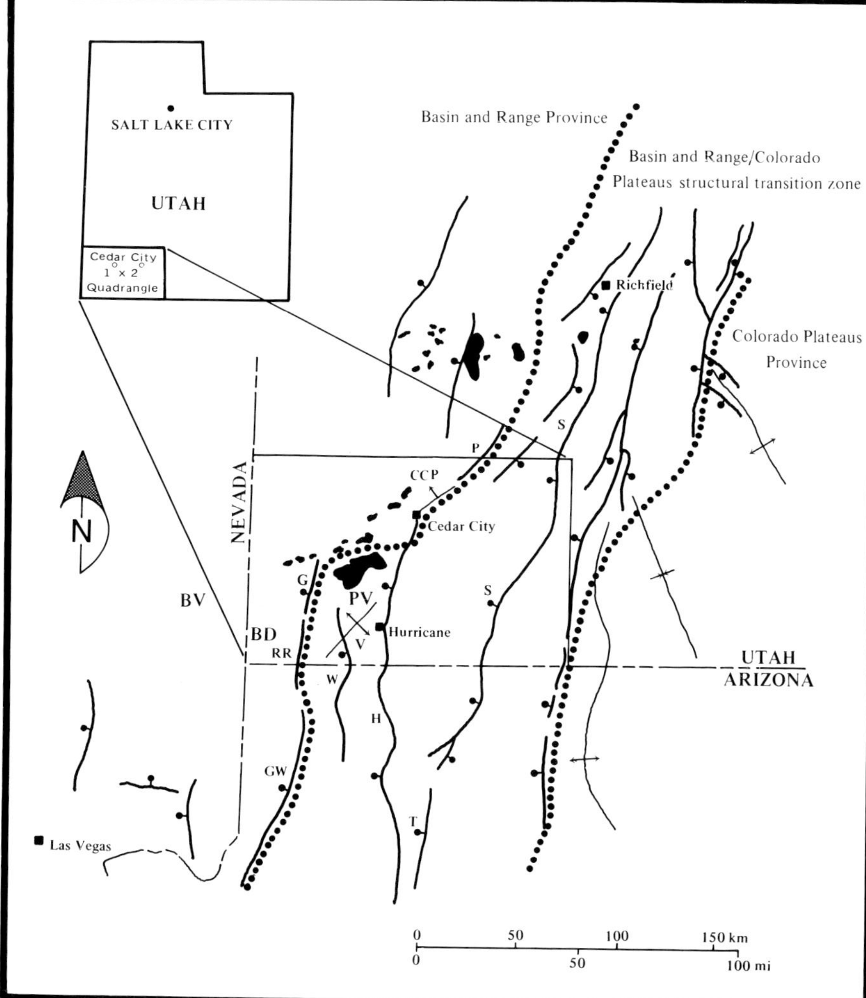

Figure 1. *Generalized index map showing location of the Cedar City 1°X 2° quadrangle relative to selected major structures, province boundaries, and physiographic features (BV, Bull Valley Mountains; BD, Beaver Dam Mountains; PV, Pine Valley Mountains) (modified from Cook, 1957). Dotted lines mark province boundaries.*

one-third is part of the Basin and Range. Between these is the structural transition zone, an area that was affected by gentle east-tilting of broad blocks bounded by widely spaced normal faults. Northeast of Cedar City the boundary between the Basin and Range and structural transition zone coincides with the base of the Hurricane Cliffs. Between Cedar City and Parowan, the main structure is a monocline (Threet, 1963) whereas northeast of Parowan it is a down-to-the-west fault. Southwest of Cedar City the location of the boundary is not so clearly defined. Many geologists have regarded the Hurricane fault in Utah as the boundary, but Anderson and Mehnert (1979) suggest, on the basis of late Cenozoic paleogeography and structure, that the boundary extends southwest from Cedar City along the north flank of the Pine Valley Mountains and from there it merges south-westward with the Gunlock fault zone which, in turn, projects southward into the Grand Wash fault in Arizona (figure 1).

The Pine Valley and Bull Valley Mountains form the eastern end of an east-trending arcuate belt of late Miocene igneous activity, and at least one older belt of similarly oriented Cenozoic igneous activity lies to the north of that area and extends eastward into the plateaus of the structural transition zone (figure 1) (Stewart and others, 1977). The Cedar City quadrangle also transects a belt trending north-northeasterly of predominantly post-Miocene basaltic volcanism (Best and Brimhall, 1974).

The intensity of late Cenozoic deformation increases from south to north and from east to west throughout the quadrangle. Therefore, equal-intensity deformational belts trend northeast. It is not known to what extent this pattern is inherited from or influenced by preexisting structural trends such as the Cordilleran hingeline (Burchfiel and Hickcox, 1972), the east margin of the Sevier orogenic belt (Armstrong, 1968), or Laramide-age folding (Cook, 1960), all of which trend northeastward through the quadrangle. Whether inherited or not the trend appears to extend into the Quaternary, as evidenced by northeast-trending belts of Quaternary faulting and basaltic vent alignments (plate 1 and Hintze, 1980) and a northeast-trending zone of post-Miocene basalts (Smith and Luedke, 1984).

SEISMICITY

The Cedar City quadrangle is at the southern end of the Intermountain seismic belt (ISB), a north-south-trending zone of seismicity extending from northwestern Montana to southwestern Utah (Smith and Sbar, 1974). The zone generally follows the eastern edge of the Basin and Range province through Utah; the part of the ISB in the Cedar City quadrangle is centered along the Hurricane Cliffs (figure 2). In the vicinity of St. George, the ISB intersects the southern Nevada seismic zone, an east-west-trending zone of relatively high seismicity (Anderson, 1978).

Figure 2 shows the locations of epicenters greater than Richter magnitude (M_L) 2.0 in the quadrangle for the period from 1850 to June 1988 plotted from the University of Utah Seismograph Stations catalog. Locations and magnitudes of earthquakes before June 1962 are approximate and based principally on noninstrumental intensity data from felt reports. Routine epicenter determinations based on instrumental data from a statewide network of seismograph stations began in July 1962, and accuracy was further improved in 1974 with full deployment of an extensive network of telemetered seismic stations (Arabasz and others, 1979).

Earthquakes are typically small to moderate in size, generally less than Richter magnitude (M_L) 4.0. Concentrations of activity are in the Cedar City/Enoch area and in the Sevier Valley area north of Panguitch, with a paucity of activity in the southeastern corner east of Kanab (figure 2). The largest earthquake in the quadrangle in historical time occurred in 1902 (estimated M_L 6.3) in Pine Valley, 33 km north of St. George. Estimated M_L5.7 and 5.5+ earthquakes occurred near Kanab in 1887 and 1959, respectively. Approximate magnitude 5 earthquakes occurred in St. George in 1891, Pine Valley in 1902 (aftershock of the earlier M_L 6.3 earthquake), Parowan in 1933, and Cedar City as a part of a swarm in 1942. The remainder of the historical earthquakes have been smaller than magnitude 5 and scattered throughout the quadrangle.

Even the largest earthquakes in the area are not of a magnitude generally thought necessary to cause surface faulting, and coseismic surface rupture has not been reported in historical time. Although some of the small to moderate earthquakes are located in the general vicinity of various Quaternary faults, the large ones are not, and this, along with the error involved in locating epicenters for older earthquakes, makes any genetic association between mapped Quaternary faults and earthquakes difficult to demonstrate (Smith, 1979). All earthquakes have shallow foci ($<$15 km) with principally dip-slip (normal faulting) focal mechanisms (Arabasz and Julander, 1986).

The occurrence of earthquake swarms in the region is common and includes swarms:(1) near Orderville in 1926-27, (2) near Panguitch in February-April 1937, (3) near Cedar City in August-September 1942, (4) in the Cedar Valley/Parowan Valley area near Enoch in November 1971, (5) near Kanarraville (two separate clusters) from December 1980 through May 1981, and (6) near Panguitch (two separate clusters) in March 1981 (figure 2) (Arabasz and Smith, 1979; Richins and others, 1981). The largest events in these swarms are in the magnitude 3 to 4 range, but two events during the 1942 swarm near Cedar City may have been in the high 4 to low 5 magnitude range. The formation of several north-trending fissures in alluvium about 24 km north of Enoch and parallel to faults in the area was noted by residents 2-3 months before the 1971 swarm. These features and an associated rise in water levels in the area, at a time when they were generally declining, may be indications of a dilatent phase prior to the swarm (Arabasz and Smith, 1979). Focal depths for earthquakes in this swarm ranged from 3 to 9 km, with normal faulting (extension direction east-northeast to west-southwest) indicated for the focal

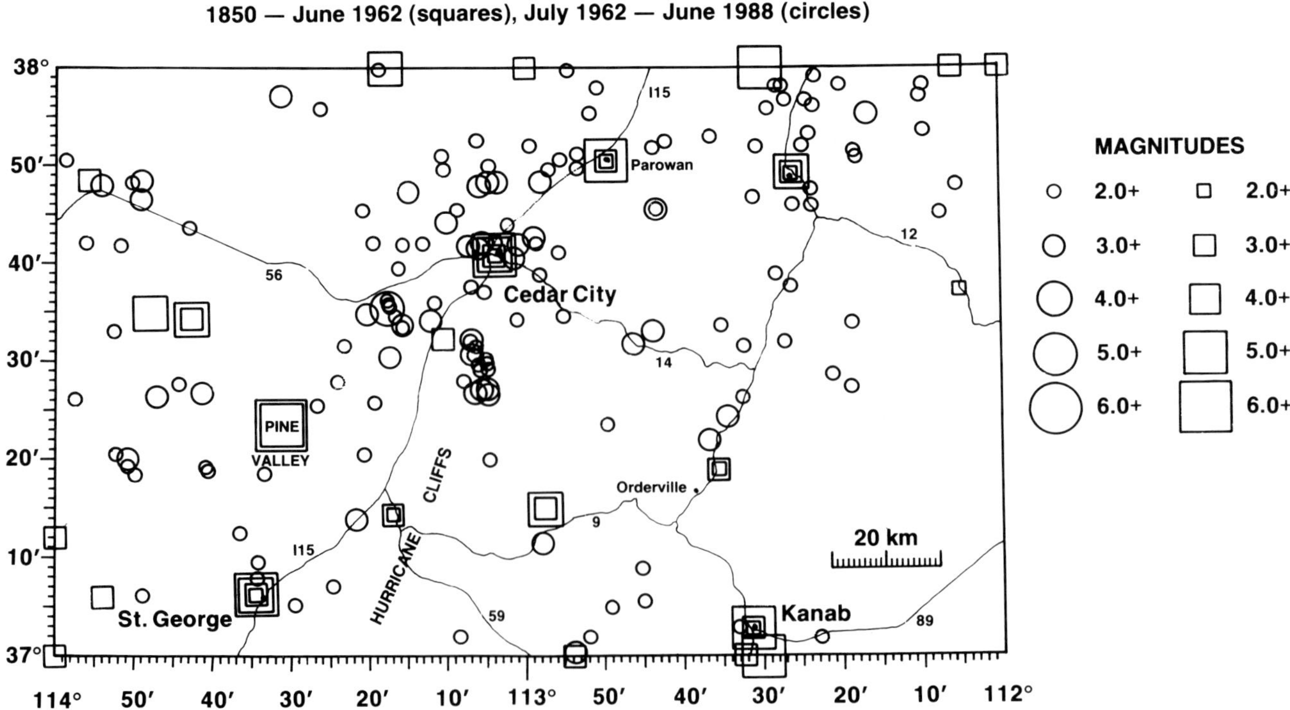

Figure 2. *Historical seismicity in the Cedar City 1°X 2° quadrangle from the University of Utah Seismograph Stations historical catalog (squares) and instrumental catalog (circles).*

mechanism of the largest earthquake in the swarm (ML 3.7) (Smith and Sbar, 1974). Focal mechanisms of earthquakes in the 1980-81 swarms near Kanarraville indicated northwest to southeast and west-northwest to east-southeast extension directions (Richins and others, 1981).

QUATERNARY HISTORY OF SOME MAJOR STRUCTURES

In this section we describe elements of Quaternary history of some range- and plateau-bounding faults and folds that are well known because they form prominent bedrock escarpments that comprise some of the beautiful scenery in the region. The largest structures represent the greatest earthquake hazard within the quadrangle because they have the greatest Quaternary displacement rates and are capable of generating the largest earthquakes. In addition, the earthquake risk along these structures is magnified by the preferential development of communities along or near them. There have not been multidisciplinary seismic hazards studies conducted in southwestern Utah similar to those along the Wasatch Front in northern Utah. As a result, our understanding of hazards related to large structures such as the Hurricane fault are sketchy by contrast.

Gravity data (Cook and Hardyman, 1967; Sawyer and Cook, 1977) and industry-acquired proprietary seismic-reflection data indicate the presence of horst blocks buried beneath the sediments of Escalante Valley and intra-valley faults beneath Cedar Valley. Such structures could have large Quaternary displacements and could be as hazardous as those bordering the valleys. Because so little is known about these faults, they are not discussed further.

GUNLOCK FAULT

The Gunlock fault (plate 1) is the north part of the Grand Wash-Reef Reservoir-Gunlock fault in Arizona and Utah. These faults separate terranes with sharply contrasting styles and amounts of internal extensional deformation and are regarded as marking the east boundary of the Basin and Range province. The amount and sense of dip separation of the Reef Reservoir-Gunlock faults varies along their trace (Cook, 1960), suggesting either a scissors motion or juxtapositioning of pre-deformed strata by a strike-slip component of motion. Stratigraphic throw is less on these faults than on the Hurricane fault 50 km to the east, probably not exceeding 300 m. Recent studies of bedrock in the area by the senior author reveal that one north-striking splay of the Gunlock fault north of Gunlock and several similarly oriented faults northwest of Gunlock are predominantly sinistral-slip faults. This evidence points to a component of strike-slip motion to explain the reversals of dip separation along the faults. Preliminary results of a detailed study in progress by Becky Hammond (Brigham Young University, oral communication, 1988)) of the fault-slip characteristics of the Reef Reservoir-Gunlock faults and of the relationships between faulting and folding indicate that strike-slip displacements of probable Neogene age are much more important than previously recognized.

A faulted basalt dated at 1.6±0.1 Ma by the K-Ar method (M.G. Best, written communication, 1985) is displaced down-to-the-west by only 8 m near Gunlock. Though this fault has clearly been active during Quaternary time, its long-term rate of normal slip appears to be very low (5 m/m.y.) compared to the 44 m/m.y. average long-term slip rate calculated for the Grand Wash fault near Grand Wash in Arizona (Hamblin and others, 1981). Its low apparent slip rate does not justify assuming a late Pleistocene last surface-faulting event. We assume the last event occurred in early to middle Pleistocene. Whether or not the Gunlock has experienced Quaternary strike-slip displacement is not known.

WASHINGTON FAULT

The Washington fault east of St. George (plate 1) is a 68-km-long down-to-the-west fault zone that extends northward into Utah from Arizona. It has an estimated maximum throw of 750 m in Arizona, and displacement decreases northward in Utah. Approximately one kilometer north of Washington the fault splays and Quaternary displacement is not evident.

The fault displaces preexisting geologic structures and has normal-drag and reverse-drag folding genetically associated with it. Drag folds are well exposed along the fault west of Warner Ridge. About 50 m of the Shinarump Member of the Chinle Formation forms the caprock of Warner Ridge beneath which about 400 m of the upper red member and Shnabkaib Member of the Moenkopi Formation are exposed. These strata have uniform N. 30° W. strikes and 5° -15° northeast dips, except in a zone a few meters to a few hundred meters wide along the north-striking Washington fault. At the fault, the dip of the strata reverses sharply to form narrow faulted anticlinal drag folds. In plan, these folds resemble the fish-hook pattern that is spectacularly developed along the Red Hollow fault south of Motoqua near the Nevada-Utah border (Hintze, 1986). Small-displacement subsidiary normal faults parallel to the main trace of the Washington fault dip east or west but the main fault is nearly vertical. Locally, the Washington fault dips steeply to the east, indicating reverse slip. Rake angles of striations range from pure dip slip to 50° S. on several small faults and the main fault, indicating a component of sinistral slip.

North of the Virgin River near Washington, the fault is at the western end of a basalt-capped ridge (Washington Black Ridge). The amount of Quaternary throw along the main fault is not known because the basalt capping the ridge is not exposed on the downthrown side, either due to removal by stream erosion or burial beneath alluvium of Mill Creek. It is also possible that the flow terminated here and never extended across the fault. However, a subsidiary fault within the Washington fault zone 0.7 km east of the main trace displaces the basalt, age-dated at about 1.1 Ma (Best and others, 1980), with a maximum throw of 4.5 m (figure 3A).

A major factor in scarp development along parts of the Washington fault is the contrast in resistance to erosion of lithologies on upthrown versus downthrown blocks. There is no definite evidence of Quaternary offset across the fault at its northern end as indicated by study of the scarp and exposures in excavations along 200 East Street in Washington. North of Washington a drainage ditch and incised channel, cut through an 8-m-high scarp, exposes faults in Mesozoic bedrock at the scarp and about 50 m to the west. Here, a Pleistocene alluvial/eolian deposit with a strongly developed but highly degraded pedogenic carbonate horizon overlies a gently dipping pediment surface cut on the Mesozoic rocks. This deposit is being eroded and undercut as the softer Mesozoic strata beneath the deposit are removed. Much of the area east of the fault is an exhumed pediment surface from which the deposit, if ever present, has been removed. Preliminary evaluation of topographic maps (contour interval 12 m) indicates the surface is approximately 70 m below adjacent 1.1 Ma basalts of Washington Black Ridge, and 25 m above the present level of Mill Creek. This indicates an average downcutting rate of 86 m/m.y., which is in general agreement with estimated average downcutting rates of 90 m/m.y. for this area during the last 2 m.y. (Hamblin and others, 1981). Assuming uniform downcutting, the approximate age of the pediment surface and deposit is perhaps 300,000 years. One remnant of this deposit is present on the downthrown side of the fault, but offset of the deposit or the erosional bedrock surface beneath it is not evident across the fault. Bedrock exposures on the downthrown side are at concordant elevations with those in the face of the fault scarp. This indicates that the scarp is primarily the result of differential erosion, accelerated on the downthrown side by ground water and more intense fault deformation of the bedrock. Apparently, little offset has occurred on this trace of the Washington fault at least during late Quaternary time. Peterson (1983) found similar relationships in his study of the Washington fault in Arizona where the major controlling factor in scarp development was differential erosion and not fault displacement.

Five exploratory trenches were excavated across the Washington fault zone in 1981-82 by Earth Science Associates (1982) as part of a seismic safety investigation of dams in the area west of Warner Ridge (locality A, plate 1). Bedrock of the Triassic Moenkopi and Chinle Formations is widely exposed in the eastern (upthrown) block and locally exposed in transverse drainages west of this part of the fault. Bedrock shows strong homoclinial flexing or normal drag in a zone as much as 300 m wide near the fault zone at the location of the trenches and for at least 3 km farther to the south. Each trench exposed alluvial deposits that were grouped by the consulting geologists into "older" and "younger" units. The excavations yielded no material for radiometric age determinations, nor were there soils suitable for relative-age characterization. The younger alluvium is offset 5 cm in one trench, less in a second trench, appears to be offset in another, and shows no offset in the other two. The small offsets of the younger alluvium may not be coseismic but could be the result of differential compaction (Earth Science Associates, 1982). However, the older alluvium in each trench is faulted. The offsets are more than one meter and probably represent one or more surface-faulting events. Unfortunately, the age of the faulted "older" alluvium is not known. Earth Science Associates (1982) estimate it to be 10,000-25,000 years old based on evidence from soil profiles exposed in trenches, but we consider the evidence to be inconclusive.

Other evidence for late Pleistocene surface faulting on the Washington fault is a remnant of a scarp (figure 3B) formed on a highly dissected pediment apron at the west base of Warner Ridge (locality B, plate 1). The pediment apron is weakly consolidated to unconsolidated, contains abundant angular cobbles and boulders, is generally less than a few meters thick, and has a well-developed desert pavement. Where exposed at the fault, angular slabs in the gravel are rotated into approxi-

Figure 3. *A) Scarp (4.5 m high) in basalt 1.1 Ma capping Washington Black Ridge south of Washington. B) View looking east-southeast at scarp separating surfaces on coarse pediment gravel (arrow marks upthrown surface that projects into camera; photographer is standing on downthrown surface covered with grass and brush).*

mate conformity with the steeply dipping fault. The surface that is offset by the fault slopes westward 4°-6°. A single profile of the scarp at the only location where it can be profiled indicates a maximum slope angle of 17° and a scarp height of 3.5 m (figure 4). These values fall close to the regression line for Bonneville shoreline scarps on a semilog plot of scarp height versus maximum scarp-slope angle, indicating that this scarp is probably of late Pleistocene age. This age is consistent with the degree of degradation of a bedrock scarp in the area that is formed on weak gypsiferous rocks of the Shnabkaib Member of the Moenkopi Formation. Those rocks probably degrade much like alluvium, and the general morphology of the scarp is suggestive of late Pleistocene surface faulting.

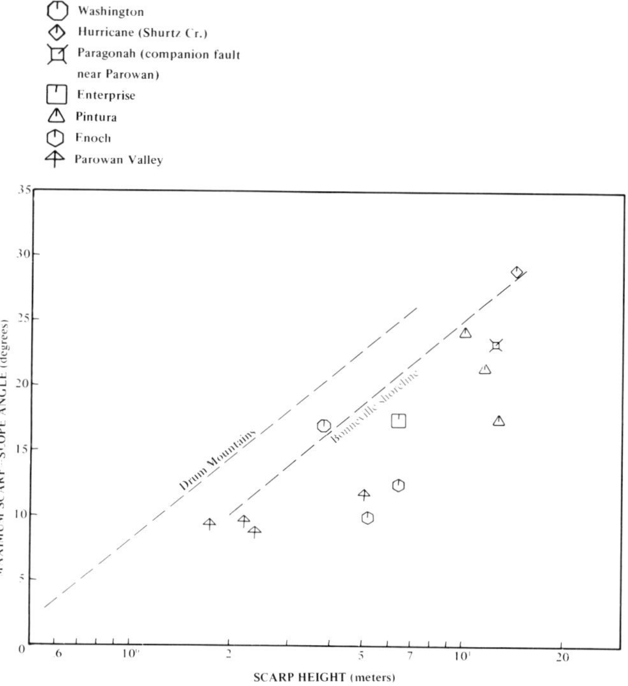

Figure 4. *Plot of maximum scarp-slope angle versus log of scarp height for selected scarps in the Cedar City quadrangle. Labeled dashed lines are reference regression lines on data from scarps of approximately 10,000 yr (Drum Mountains) and 15,000 yr (Bonneville shoreline) from Bucknam and Anderson (1979).*

In summary, we map a 4-km-long part of the Washington fault along Warner Ridge as late Pleistocene and the parts passing through Washington and into Arizona as middle to late Pleistocene. Scarborough and others (1986) place the Washington fault in Arizona and southern Utah in a category of faults displacing latest Pleistocene or Holocene alluvial deposits or with alluvial fault-scarp morphologies indicating an age of less than 30,000 years. More detailed study by Menges and Pearthree (1983) shows that scarp morphologies in southern Utah indicate uncertainty in the age ranging from early Holocene to late Pleistocene; middle to late Holocene deposits are not displaced. This assessment is in general agreement with ours, although our expanded age range of middle to late Pleistocene indicates greater uncertainty.

Several north-trending normal faults parallel to the Washington fault are found in St. George. One trends along the east side of the city (Christenson and Deen, 1983). Another is exposed in the Red Hills north of the St. George City Office Building and projects southward just east of the St. George Mormon Temple. Although expressed in bedrock north of St. George, these faults become obscure to the south. No evidence for Quaternary movement was found on either of these faults, although both have been modified by development and scarps may have been leveled and other surficial evidence destroyed. Neither fault is shown on the map.

ANTELOPE RANGE FAULT

The range-bounding fault along the west flank of the Antelope Range forms the southeast edge of the Escalante Desert and is a northeast-trending Basin-and-Range structure active during Quaternary time. The area near the town of Newcastle at the south end of the fault is a known geothermal resource area. The fault, as shown, is taken from mapping by Shubat and Siders (1988), Siders and others (1988), and Grant and Proctor (1988). It consists generally of a single trace along the bedrock-alluvium contact at the base of the range. Two prominent right steps occur in the fault trace, one in the Newcastle area and the other at the north end of the range. A complex zone of faulting is present within the zone between the faults at the right step near Newcastle.

The section of the fault with Quaternary movement extends for about 23 km along the range front. Triangular facets, a steep linear bedrock front with a distinct steepened segment at the base, and active alluvial-fan deposition at the mountain front all indicate a tectonically active range front. Evidence for Quaternary faulting at the front ceases south of Newcastle, except for a few widely spaced, short, faceted spurs (unmapped).

Scarps on alluvium are present at the north end of the range and in a few isolated places along the range front. These scarps were not profiled, but approximate measurements of scarp heights and maximum slope angles were made in the field. Scarps at the north end of the range have maximum slope angles of less than 5° and appear to be older than most scarps on alluvium at the range front which are nearly the same height (20-30 m) but are steeper (average 15°). These values are not shown in figure 4 because they are approximate, but they suggest an age greater than Bonneville shoreline scarps. The heights of these scarps on alluvium indicate recurrent movement during Pleistocene time. The most recent displacements have not recurred along some fault traces, as evidenced by exposures of faults lacking scarps in dissected middle to late Pleistocene alluvium in the area of the right step near Newcastle.

Age and displacement data for offset alluvial surfaces are insufficient to determine slip rates or approximate timing of events. However, younger alluvial-fan surfaces of Holocene age are not offset and scarps in Pleistocene alluvium are highly degraded, indicating a probable elapsed time since the last event of well over 10,000 years.

HURRICANE FAULT

An understanding of the seismotectonics of the Hurricane fault is important to the evaluation of seismic hazards in the relatively small but growing population centers such as Cedar City and St. George that are located on or near the fault. The general features of the Hurricane fault are described by Hamblin (1970a). In detail, the fault has a conspicuous dog-leg trace. Locally the fault is a zone ranging to more than 1 km wide. In Utah, the fault follows preexisting northeast-trending folds of probable Laramide age over long distances. Published estimates of normal separation on the fault in Utah range over almost an order of magnitude from 430 to 4000 m (Anderson, 1980). The great discrepancy arises, in large part, from the failure of several investigators to subtract from the total apparent throw: 1) prefault folding of possible Laramide age, 2) reverse-drag flexing of the hanging wall, and 3) rise-to-the-fault flexing of the footwall. Using unpublished geologic map compilations provided to us at 1:62,500 scale by W.K. Hamblin for the St. George and Hurricane quadrangles and by E.G. Sable at 1:100,000 scale for the St. George quadrangle, we constructed apparent dip components using 3-point solutions at distances sufficiently far from the fault zone to be representative of the block interiors, projected those to the fault, and measured the throw. Tectonic displacements (Swan and others, 1980) of 1100 and 1500 m are obtained at the latitudes of St. George and Toquerville, respectively. It is doubtful that the tectonic displacement or throw exceeds 2 km anywhere on the Hurricane fault.

Several lavas ranging in age from a few thousand years to several million years have flowed across the fault in Utah and Arizona providing an unusually good record of recurrent movement (Hamblin, 1970b; Hamblin and others, 1981). Of special significance to the present study are those with ages of about 1 Ma and younger. Fortunately these flows are widespread along the part of the Hurricane fault in Utah and its coextensive structures to the northeast—the Cedar City-Parowan monocline and Paragonah fault. At some localities individual flows or series of flows can be found on both sides of the main structures, thus providing an ideal situation for estimating vertical stratigraphic offset and slip rates. A summary of available data is given in table 1. From these data the stratigraphic throw on the Hurricane fault in Utah ranges from 300 to 470 m in the last million years. These values are probably maximum values and exceed the tectonic throw that results from subtracting fault-related flexing. If, as is suggested above, the total tectonic throw that must be accounted for is 1100-1500 m and if a uniform adjusted tectonic displacement rate of 300 m/m.y. is assumed, the throw on the Hurricane fault is mainly Pliocene and Pleistocene.

Averitt (1962) described and illustrated a conspicuous scarp on the principal range-front strand of the fault at the mouth of Shurtz Creek about 8 km south-southwest of Cedar City. The scarp is at least 20 m high, is formed on very coarse bouldery alluvium, has a maximum slope angle of 20°, and is deeply incised by Shurtz Creek. On the basis of its profile, the age of most recent displacement appears to be pre-Holocene but is probably close to the Pleistocene-Holocene boundary (figure 4). The surface that is displaced at this scarp was referred to by Averitt (1962) as the Shurtz Creek pediment. The Quaternary gravels that mantle the pediment are parent material for a soil that is definitely pre-Holocene and is probably more than 50,000 years old (Anderson, 1980). In profile, the scarp shows no clear evidence of recurrent movement, so it is unclear how many surface-faulting events are represented by its 20-m height. Assuming a minimum age of 50,000 years for the surface, a maximum displacement rate of 400 m/m.y. is indicated.

Three kinds of geomorphic features at the base of the Hurricane Cliffs, in addition to the Shurtz Creek scarp, indicate late Pleistocene or younger surface displacement. First, between Cedar City and the latitude of New Harmony there are several short sections where the cliff base is not only steep but is formed on relatively nonresistant claystones and evaporite-bearing siltstones and mudstones of the Chinle and Moenkopi Formations (Averitt, 1962). Some of these rocks are probably less resistant to scarp degradation processes than the boulder-rich pediment gravels from which a close-packed lag mantle has developed on the scarp face at Shurtz Creek. Second, there are several places where the cliff base is formed on resistant upper Paleozoic carbonate rocks. Where small and intermediate transverse ephemeral streams cross the cliff base on the resistant rocks, nick points are so sharp that storm-generated runoff in some of these transverse drainages produces spectacular waterfalls at the mountain front. Third, small areas of pedimented bedrock along minor transverse drainages are truncated sharply at the cliff base. Near Pintura (figure 5), the upthrown and downthrown parts of the offset pediment surface are preserved and are separated by a precipitous bedrock scarp. Together with the scarp at Shurtz Creek, these three kinds of features constitute strong evidence for a substantial rate of late Pleistocene surface offset on the Utah

Table 1.

Displacement rates estimated for the Paragonah fault, Cedar City-Parowan monocline, and Hurricane fault on the basis of stratigraphic offset of basalt flows and for the Wasatch fault on the basis of fission-track ages from the Precambrian Farmington Canyon Complex of the Wasatch Mountains.

Area	Rate (m/m.y.)	Basis	Reference
Paragonah fault, main splay	295	K-Ar age of 0.44±0.04.	Fleck and others, 1975; Hamblin and others, 1981.
Paragonah fault, main splay plus subsidiary fault.	455	K-Ar age of 0.44±0.04.	Fleck and others, 1975; Hamblin and others, 1981.
Cedar City-Parowan monocline	250	K-Ar ages of 1.0±0.16, 0.93±0.14, and 1.1±0.1 from Cinder Hill and Braffits Creek.	Anderson and Bucknam, 1979a; Anderson and Mehnert, 1979.
Hurricane fault at Hurricane	300	Six K-Ar and one thermo-luminescence determinations.	W.K. Hamblin and M.G. Best, written communications, 1979; Hamblin and others, 1981.
Hurricane fault at Pintura	470	Unpublished K-Ar age of about 1.4 m.y.	M.G. Best, oral communication, 1979.
Hurricane fault at North Hills	400	K-Ar ages of 1.09±0.34 and 1.06±0.28 on offset basalt.	Anderson and Mehnert, 1979.
Southern Hurricane fault, Arizona	123-250	Thermoluminescence analysis of lava flows.	W.K. Hamblin and M.G. Best, written communication, 1979.
Bountiful-Ogden (Wasatch fault).	400	Forty fission-track ages on apatite.	Naeser and others, 1980.

Figure 5. *View to the east from directly north of the town of Pintura showing the Hurricane Cliffs composed of gently east-dipping upper Paleozoic rocks and a small transverse drainage issuing from the cliffs. Double-shaft arrows mark the main fault trace. Where the transverse drainage crosses the main fault trace, a steep bedrock cliff separates remnants of the apical part of a pedimented surface (marked by single-shaft arrow) from its downdropped equivalent (surface covered by juniper trees).*

part of the Hurricane fault. We emphasize that Holocene surface displacement is undocumented but certainly not precluded.

Total displacement on the Hurricane fault decreases southward into Arizona (Hamblin, 1965). Hamblin and others (1981) report that the average slip rate for the Hurricane fault near Hurricane is 300 m/m.y. In Arizona near where the fault crosses the Colorado River, there is evidence of late Pleistocene displacement with a rate less than that of the northern Hurricane fault. M.G. Best (written communication, 1987) reports 22- and 25-m offsets of basalts age dated at two separate localities by the thermoluminescence method (Holmes, 1979) at 88 ± 15 Ka and 203 ± 24 Ka respectively. These yielded displacement rates of 250 m/m.y. and 123 m/m.y., respectively.

Though the Hurricane and Wasatch faults contrast sharply in several important aspects as summarized by Anderson and Mehnert (1979), long-term vertical displacement rates on the Hurricane fault are comparable to or within a factor of 2 of those determined by Naeser and others (1980) for the Bountiful-Ogden section of the Wasatch fault calculated from fission-track analyses of apatite (table 1). Continuous youthful-looking scarps on alluvium such as those that are common along major segments of the Wasatch fault are not found along the Hurricane fault. Alluvial-fan surfaces adjacent to the bedrock base of the Hurricane Cliffs do not appear to be displaced. Most of the fans are small and are probably of Holocene age. The absence of a clear record of Holocene displacement on the Hurricane fault suggests several possibilities:

1. Large displacements have occurred during the Holocene but have not left an obvious geomorphic record.
2. The fault cannot withstand large shear stresses, and therefore moves in very small increments that do not result in conspicuous scarps.
3. Strain has been stored on the fault at a rate consistent with the long-term average, but the return period for strain release is somewhat longer than the Holocene Epoch.

We comment on these possibilities in turn. There does not appear to be any process along the Hurricane Cliffs that would serve to systematically remove evidence of significant Holocene surface faulting. Rates of sedimentation and erosion cannot be inferred to be especially high as compared, for example, to those along the Wasatch Front. Therefore, the first possibility does not seem reasonable. The second possibility seems unreasonable in light of the 20-m-high scarp at the mouth of Shurtz Creek. The third possibility is preferred.

The elapsed time since the last surface-faulting earthquake on the entire 80-km-long part of the Hurricane fault in Utah is probably somewhat greater than 10,000 years. Evidence regarding average recurrence and displacement per event is lacking and cannot be reliably estimated from calculated long-term displacement rates.

PARAGONAH FAULT

The Paragonah fault is the easternmost of a system of north- to northeast-striking faults that splinter the northwest-vergent Cedar City-Parowan monocline (Threet, 1963). The monocline forms a structural bridge between the Paragonah and Hurricane faults. North of Paragonah, the fault is at the base of the Hurricane Cliffs where it marks the east boundary of the Basin and Range. South of Paragonah, the fault enters the Hurricane Cliffs and penetrates the edge of the Markagunt Plateau (figure 6). Unlike the Hurricane fault, the Paragonah fault adjacent to Parowan Valley is not complicated by preexisting folds along its trace. The upthrown (eastern) block exposes gently north-northeast-tilted Upper Cretaceous and lower Tertiary sedimentary strata that are broken cleanly at the fault with only very minor normal drag. The bedrock scarp is straight and precipitous, indicating recurrent youthful displacement.

A basalt flow dated by Fleck and others (1975) as 0.44 ± 0.04 Ma was extruded from the Hurricane Cliffs southeast of Paragonah and flowed down ancestral Water Canyon across the Paragonah fault and a companion fault to the west. The two faults displace the basalt with a combined throw of about 200 m. Hamblin and others (1981) report a displacement rate of 295 m/m.y. for the main strand of the Paragonah fault, and offset of the basalt by the companion fault yields a displacement rate of about 160 m/m.y., for a cumulative rate of about 455 m/m.y. at the plateau margin (table 1). The age of the last displacement on either fault is not known. Scarps on alluvium are not recognized along the Paragonah fault, but a poorly preserved scarp formed on boulder gravel along the trace of the companion fault directly east of Parowan was profiled (figure 4). The scarp is about 12.5 m high and may represent several surface-faulting events. Its maximum slope angle of 23.3° would suggest that the age of the youngest event is latest Pleistocene. Available data can be interpreted in a number of different ways, all of which indicate late Pleistocene surface faulting at a rate analogous to that of the Hurricane fault.

CEDAR CITY-PAROWAN MONOCLINE

On the basis of age-dated basalts, Anderson and Bucknam (1979b) presented geologic evidence for at least 250 m of uplift in the last 1 m.y. as a result of folding in the central part of the Cedar City-Parowan monocline (table 1). If the monocline is a structural bridge between the Hurricane and Paragonah faults, as suggested by Threet (1963), this minimum uplift rate is broadly consistent with maximum rates of 455 m/m.y. for the combined displacement on the two faults at Paragonah and 320 ± 80 m/m.y. estimated for the northern Hurricane fault.

The Cedar City-Parowan monocline has not been mapped in detail, but scattered field investigations suggest that it is a very complex structure involving a complex stratigraphic assemblage. A normal stratigraphic sequence consisting of Upper Cretaceous and lower Tertiary sedimentary rocks overlain by middle to upper Tertiary volcanic rocks (Threet, 1963; Anderson and others, 1975) is, in places, structurally attenuated and is generally overlain by a very complex assemblage of locally derived, commonly monolithic debris flows and landslides. In addition to being monoclinally flexed and splintered by several north- to northeast-striking normal faults of large displacement, this entire stratal assemblage is deformed both by strike-slip faults and interrelated systems of map-scale to mesoscopic folds and faults that tend to displace rocks down toward the uplifted part of the monocline. The down-to-the-mountain folding and faulting has produced numerous closed basins perched along the mountain front (figure 7). These range in size from a few hundred square meters to 1 km². That they are only partially filled with sediment in a mountain-flank physiographic setting suggests youthful deformation. Folds and faults associated with the largest one (Elliker basin, locality C, plate 1) involve basalt that yielded a K-Ar age of 0.9 ± 0.14 Ma (Anderson and Bucknam, 1979b). The style of folding and faulting responsible for producing these features is analogous to that seen in excavated cinder beds at Cinder Hill (figure 8) where the structures have extremely limited depth penetration. Even the largest down-to-mountain faults (not shown on plate 1), such as those at Elliker basin, are not likely

Figure 6. *View to the south from Interstate 15 of the Paragonah fault forming the east edge of Parowan Valley. Note steep, linear mountain front and triangular facets indicating youthful displacement. The fault diverges from the mountain front and trends into the Markagunt Plateau to the south, shown in the right side of the figure.*

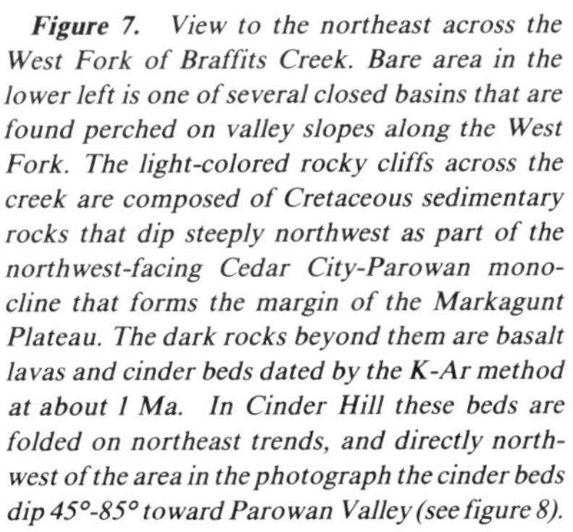

Figure 7. *View to the northeast across the West Fork of Braffits Creek. Bare area in the lower left is one of several closed basins that are found perched on valley slopes along the West Fork. The light-colored rocky cliffs across the creek are composed of Cretaceous sedimentary rocks that dip steeply northwest as part of the northwest-facing Cedar City-Parowan monocline that forms the margin of the Markagunt Plateau. The dark rocks beyond them are basalt lavas and cinder beds dated by the K-Ar method at about 1 Ma. In Cinder Hill these beds are folded on northeast trends, and directly northwest of the area in the photograph the cinder beds dip 45°-85° toward Parowan Valley (see figure 8).*

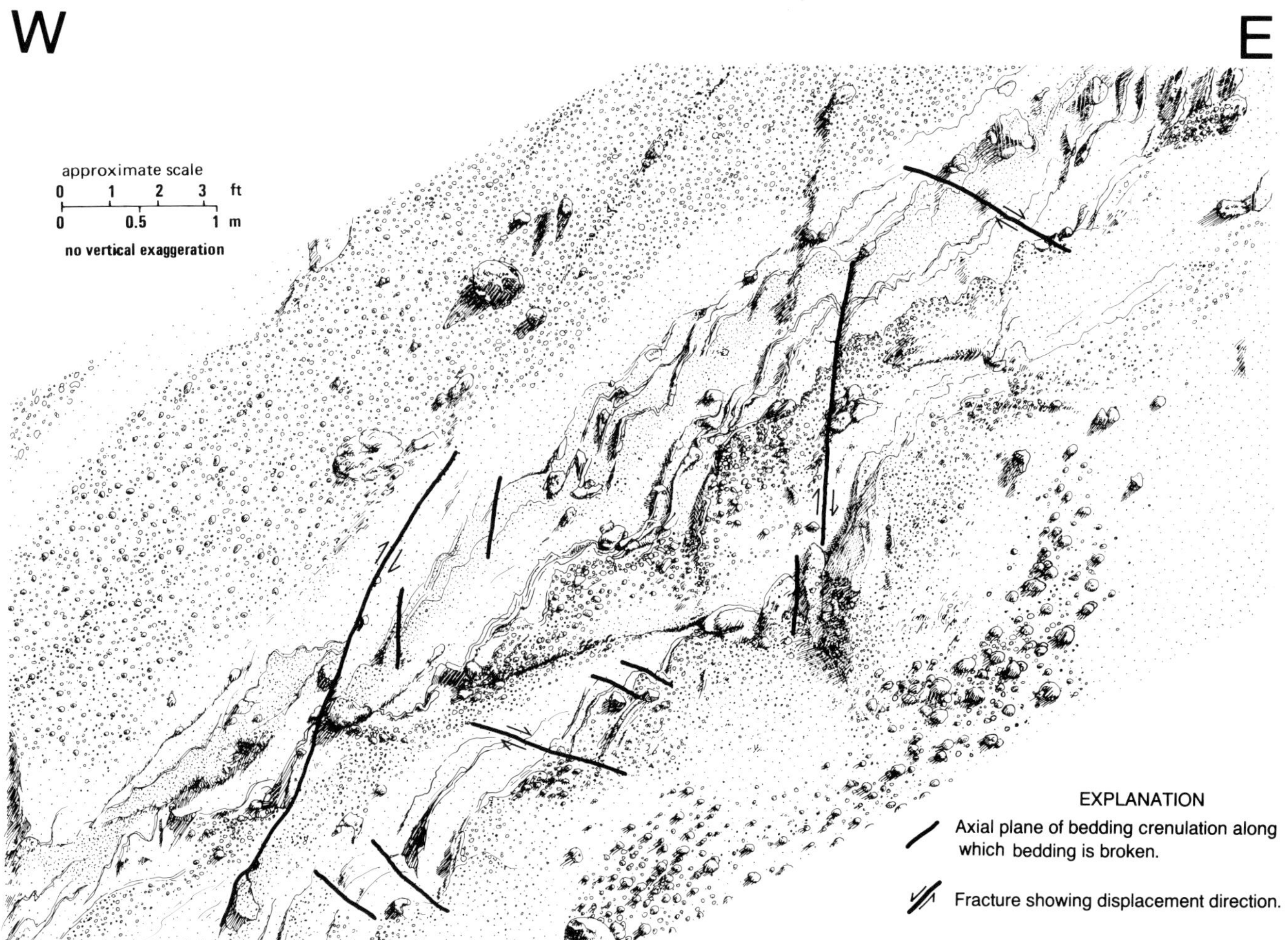

Figure 8. *North-looking sketch of excavation in poorly consolidated, bedded cinder deposits at Cinder Hill showing details of crenulated bedding in an interval of fluvial(?) beds sandwiched between massive air-fall beds. The westerly dip of the fluvial(?) beds is about 45°, but parts of the beds between crenulations dip much steeper—locally as much as 85°. Bedding is broken along the axial plane of some crenulations. The axial planes and fractures can be grouped into steeply and gently dipping systems both of which represent components of down-to-the-east displacement. The style of deformation illustrated here is important because it is analagous to larger scale folds and faults at Cinder Hill, Elliker basin, and at several other locations in the Braffits Creek area. The deformational style shown here is common in neotectonic monoclines elsewhere in southwest Utah (Anderson and Barnhard, 1987). It shares many features with "ductile faulting" as described from competent crystalline rocks by Davis (1980). Here, ductile deformation is accomplished by grain-boundary translation and rotation in poorly consolidated materials. There is an inverse correlation between bed thickness and apparent strain magnitude.*

to penetrate more than a kilometer or so. We therefore defer discussion of these features to the section on nonseismogenic structures. The possibility exists, however, that the main mountain-front monocline is the complex surface expression of a blind plateau-bounding normal fault zone that has seismic potential equal to that of other major faults in the region. To test this possibility requires subsurface exploration.

RED HILLS FAULT

The Red Hills are separated from Parowan Valley by the Red Hills fault. The hills are composed principally of Tertiary Claron Formation. The southern part of the faulted front near Little Salt Lake in Parowan Valley is tectonically active as indicated by the linear bedrock-alluvium contact and steep, active alluvial fans with apices at the front. Very few fans are entrenched at the fanhead and, in those that are, the locus of

deposition has shifted down-fan less than 0.4 km. Despite the relatively nonresistant bedrock at the front, few embayments occur even along major drainages. Also, the front is marked by a steep set of triangular facets at the base of the slope, indicating that the front has been in a period of active uplift during at least late Pleistocene time. Few alluvial scarps are present along the fault, chiefly because fan apices are at the mountain front and do not extend upstream across the fault. Isolated short scarps of possible fault origin are on colluvium, and one (0.2 km long) is found on late Pleistocene alluvial-fan deposits near the southern end of the fault.

The presence of Little Salt Lake and the broad, flat-bottomed Parowan Valley is an indication of youthful activity along the Red Hills fault. The outlet of the lake is at the head of a canyon (Parowan Gap, locality D, plate 1; cover) cut through the Red Hills as they rose as a horst. During late Pleistocene time, it is probable that subsidence of the basin accompanied uplift of the Red Hills, therefore maintaining the Little Salt Lake area as a basin-floor depocenter. During Holocene time, the level of the outlet and Little Salt Lake were controlled by deposition on the Whitney Canyon alluvial fan which has partially blocked the outlet stream through Parowan Gap (Nielson, 1983). Prior to removal by man, this fan maintained 3 m of closure in the Little Salt Lake basin before overflow through Parowan Gap occurred (Nielson, 1983).

SEVIER FAULT

The Sevier fault zone, as used herein, forms the west boundary of: (1) the Sevier Plateau (mainly in Piute County north of the Cedar City quadrangle), (2) the Paunsaugunt Plateau (mainly in Garfield County), and (3) the structural block containing Bryce Canyon, the Pink Cliffs, and the White Cliffs (mainly in Kane County). The fault zone in Utah is the northern structural element in a 500-km-long fault zone that includes the Toroweap (figure 1) and Aubray faults in Arizona (Hamblin, 1970a). In the Cedar City quadrangle, the zone strikes north-northeast and consists of numerous right- and left-stepping generally *en echelon* faults on which rocks are typically downthrown to the west-northwest (Hintze, 1980). R.W. Krantz (written communication, 1985) noted a consistent southerly component to the rake of striae on the Sevier fault and other faults of the transition zone, suggesting that they are sinistral-normal faults. Displacement on the Sevier fault generally increases northward. On the basis of three-point solutions and projections to the fault, we estimate 475 m of throw at the Coral Pink Sand Dunes directly north of the Arizona-Utah border. Seismic-reflection data (E. Lundin, written communication, 1988) indicate about 900 m of throw on the basement at Red Canyon east of Panguitch. Total throw is as much as 1.5 km in southern Sevier County to the north of the quadrangle. The amount of displacement differs greatly between individual *en echelon* strands as a result of transfer of displacement from one to the other.

Throughout most of its length in the Cedar City quadrangle, the escarpment on the upthrown block of the Sevier fault is on bedrock, and along much of its length bedrock is also exposed in the downdropped block. Therefore, the opportunity to assess the Quaternary-faulting history by the study of Quaternary deposits is limited. Except where the fault cuts age-dated Quaternary basalt, age evidence for surface faulting is indirect. Our discussion of evidence for Quaternary displacement is organized geographically from south to north, beginning with a brief account of the Toroweap fault in Arizona. Because the Sevier fault is broken into *en echelon* parts on a rather small scale (Hintze, 1980; Doelling and Davis, 1989), we do not attempt to relate this segmentation to seismogenic behavior.

Though it is well known that the majority of throw on the Toroweap fault predates basaltic eruptions of the western Grand Canyon region and that progressively younger volcanics are displaced progressively less (see Koons, 1945, for a summary), available age data do not span a wide range of ages and thus cannot yield conclusive long-term displacement rates. Age determinations using the thermoluminescence method on samples of two widely separated flows cut by the Toroweap fault (Holmes, 1979; M.G. Best, written communication, 1987) yield rates of 74 m/m.y. (Volcans Throne) and 60 m/m.y. (Heaton Knolls) on rocks that yielded ages of 201 ± 34 Ka and 284 ± 48 Ka., respectively. If these data are accepted at face value, there has certainly been middle to late

Pleistocene surface faulting on the Toroweap fault. North of these basalts in Arizona, Scarborough and others (1986) indicate possible late Pliocene and Quaternary movement with 2-3 m of displacement of Quaternary units at a point about 16 km south of the Utah border. Menges and Pearthree (1983) indicate that the fault in southern Utah lacks specific stratigraphic evidence for Quaternary displacement, and the age of last surface faulting should accordingly be considered of questionable Quaternary age.

Along the part of the Sevier fault from the Utah line north to Orderville, two geomorphic features provide indirect evidence of Pleistocene deformation. First, in several places the west-facing bedrock scarp is conspicuously compound in cross profile with a moderately sloping rounded upper part and a steep cliff-forming lower part. Near Yellowjacket Spring and elsewhere, the lower cliff-forming part may result from youthful throw on the fault rather than from erosional sapping and spalling at the cliff base or differential resistance to erosion of rocks in the lower cliff. Second, the main trace of the Sevier fault has a conspicuous left step about 5 km south-southeast of Mount Carmel Junction. We measured striations on several steep small-displacement fault surfaces along the main trace south of Clay Flat (locality E, plate 1) and found a consistent southerly rake component (figure 9) similar to that reported by R.W. Krantz (written communication, 1985). The indicated component of sinistral slip combined with the left step at Clay Flats should, according to faulting models proposed by Segall and Pollard (1980) and Sibson (1985), result in an area of concentrated dilatation in the vicinity of Clay Flat. Clay Flat is a closed-basin depocenter of less than 1 km² that interrupts the transport of fluvial sediments carried north and west by ephemeral streams that drain an area of approximately 70 km² (plate 1). In order to maintain such a depocenter, active late Pleistocene subsidence seems to be required. The hazards implications of such subsidence are uncertain because the deformation could either be directly linked to surface-faulting earthquakes on the Sevier fault or to a more continuous type of low-level seismicity (or aseismic creep) distributed through a large volume of rock.

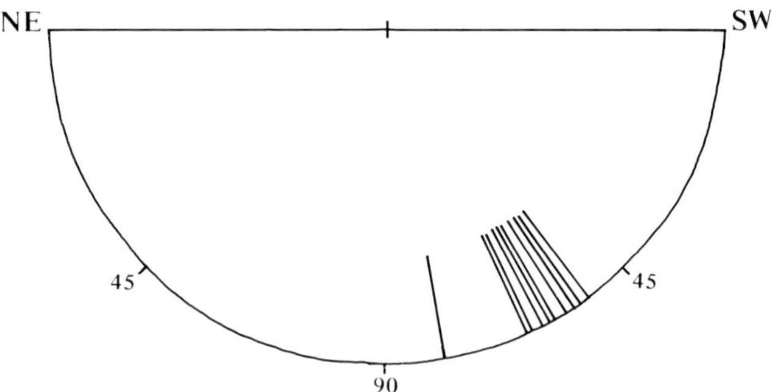

Figure 9. *Diagram showing striations measured on 10 steep minor faults in the Sevier fault zone projected onto a vertical plane striking N. 22° E. The projection plane represents the main strand of the Sevier fault. Note consistent southerly rake component with concentration between 50° and 65°. Data are from locality marked by E on plate 1.*

Spectacular bedrock scarps along the Sevier fault in the Orderville-Glendale area are largely the result of erosion of the relatively weak Cretaceous rocks that are downfaulted against the resistant Navajo Sandstone (Cashion, 1967). The precipitousness of these fault-line scarps cannot be taken as evidence for Quaternary-age displacements. At Black Mountain (8 km northeast of Glendale) a basaltic volcanic center of Quaternary age is located on or near the main trace of the Sevier fault. Our reconnaissance field investigation indicates that at least three flows issued from the center and flowed westward across the escarpment at the fault. A west-facing scarp that is conspicuous on air photos is the distal front of a flow that rests on an older flow for which Best and others (1980) report a K-Ar age of about 0.56 Ma. Neither flow appears to be faulted at the scarp, despite the coincidence of the scarp with the trace of the westernmost fault of the Sevier fault zone mapped by Doelling and Davis (1989). Other conspicuous escarpments on basalt in the area appear to be headwall scarps in landslides. Cashion (1961) indicates 23 m of displacement in basalt at Black Mountain. Though it is probable that basalt is displaced at the main trace, there is considerable uncertainty as to which scarps are related to landslides, preflow topography, or faults. There is also uncertainty as to which of the scarp-forming basalts along the main trace correlates with the age-analyzed flow. These uncertainties preclude estimating long-term displacement rates at Black Mountain from available data.

The physiographic expression of the Sevier fault in the area north and northeast of Alton is unique. A 6-km-long section of the fault there is expressed as a southeast-facing 150 to 200-m-high obsequent fault-line scarp rather than as the normal northwest- to west-facing fault scarp or fault-line scarp. Though we have evidence for surface displacement of probable Holocene age along this part of the fault, we interpret the displacement to be nonseismogenic and defer discussion to the section dealing with inferred aseismic deformation.

Where the Sevier fault forms the west edge of the Paunsaugunt Plateau in Garfield County, it marks the contact between the Cretaceous and Tertiary rocks of the Plateau and the Quaternary alluvial deposits of the Sevier River Valley. The Plateau margin is linear, but there are no known sites where the fault displaces unconsolidated alluvial deposits of known Quaternary age along the front. The only offset Quaternary deposits occur at Red Canyon southeast of Panguitch where 0.56 Ma basalts (Best and others, 1980) are offset 200 m (Anderson and others, 1978). We infer that most of the 200 m of displacement is distributed throughout the middle and late Pleistocene. North of Red Canyon, the main branch of the fault trends into the volcanic bedrock of the Sevier Plateau and dies out (Doelling, 1975; Hintze, 1980). Carpenter and others (1967) show numerous faults dashed through a mass of Quaternary rubble in this area (locality F, plate 1). These faults are shown on the Quaternary fault map of Utah (Anderson and Miller, 1979). Neither these faults nor the rubble are shown on geologic maps at the same scale (Doelling, 1975) or smaller scales (Hintze, 1980). On the basis of our study of 1:60,000-scale air photos, the rubble appears to include landslides, pediment deposits, and alluvium lacking through-going scarps, although the topography is very irregular. The fault traces are visible only as aligned drainages in bedrock, and we do not include them on plate 1.

Hintze (1980) shows a north-trending fault trace directly west of where the main trace of the Sevier fault trends into bedrock and dies out. There the plateau margin is conspicuously linear with transverse interfluves commonly terminated at triangular facets. Active alluvial-fan deposition at the mountain front, together with the plateau-margin morphology, suggest youthful fault deformation, despite an absence of

scarps on alluvium. We map this as a fault with suspected middle- to late-Pleistocene surface offset (plate 1). It may represent an *en echelon* left step in the main fault system, although the evidence for youthful displacement on the Sevier fault in the Cedar City quadrangle appears to die out north of Sanford Creek.

SCARPS ON ALLUVIUM

Some scarps on alluvium in the Cedar City quadrangle are located along major structures such as the Washington fault at Warner Ridge, the Hurricane fault at Shurtz Creek, and the Antelope Range fault. These are described in preceeding sections. Most of the remaining scarps on alluvium are described in this section by geographic area. They occupy diverse structural settings such as: (1) mid-valley faults probably located above buried horst-and-graben structures (Escalante Valley and Parowan Valley), (2) faults antithetic to the Hurricane fault (Pintura area), (3) faults that cut the Sevier Plateau transversely (Hunt Creek), and (4) faulted folds developed on the downthrown block of the Sevier fault (Race Hollow). Most of the scarps are on alluvial-fan gravels. However, those near Pintura are on very coarse debris-flow deposits and scarps at Enoch are mostly on fine-grained alluvium consisting of sand, pebbly sand, silt, and clay.

ESCALANTE DESERT

Numerous short, isolated scarps on alluvium are present in the Escalante Desert. Most trend northeast, parallel to the axis of the valley, and are accompanied by clusters of subparallel vegetation lineaments (Ertec Western, Inc., 1981). Both the principal drainage and wind-aligned features (dunes, deflation basins) parallel this northeast trend, resulting in a wide variety of northeast-trending lineaments, many of which are not related to faulting. For this compilation, only topographic scarps which cannot reasonably be attributed to water or wind action are shown. All such features are present on deposits of middle to late Pleistocene age, and none appear to displace younger (Holocene) alluvium, playa deposits, or eolian deposits. The most prominent scarps are near Enterprise, west of Newcastle, northwest of Table Butte, and along the west side of the Bald Hills and Mud Spring Hills on the east side of the valley.

The faults southeast and east of Enterprise truncate alluvial-fan deposits derived from the Pine Valley Mountains to the south. Though a poorly preserved scarp on alluvium is found directly southeast of Enterprise (figure 4), the evidence for faulting consists chiefly of concordant faceted spurs at the ends of finger-like ridges formed on dissected alluvium. The spurs are aligned oblique to modern drainage. On the basis of their discontinuous and highly modified geomorphic expression, the last faulting event is assumed to be middle to late Pleistocene in age.

The fault near Newcastle displaces deposits down to the east and may be an antithetic fault associated with the Antelope Range fault, forming the west edge of a graben along the east side of the Escalante Valley. The fault scarp is rounded with an average maximum slope angle of 5° and height of about 2 m. Because it is an uphill-facing scarp on an alluvial surface sloping less than one degree, it is not appropriate to estimate age based on relationships between scarp height and maximum slope angle. The scarp does not cut Holocene deposits and is found only on alluvial-fan deposits of Pleistocene age (Siders and others, 1989). The last displacement is conservatively estimated to be middle to late Pleistocene.

Northwest of Table Butte in the north-central part of the Escalante Desert near Zane, down-to-the-west faults are found in the valley bottom and down-to-the-east faults are on the alluvial piedmont along the west side of the valley northeast of Zane. The more northerly down-to-the-east scarps are the most prominent and cut probable middle to late Pleistocene alluvial fans formed from detritus carried from hills to the west of the scarps. These scarps are the southern part of a 19-km-long zone of scarps described by Anderson and Bucknam (1979a). They are highly dissected with apices of well-developed, post-faulting alluvial fans at their bases. North of the Cedar City quadrangle, poorly defined shoreline features that appear to represent the high stand of Lake Bonneville are developed on a post-faulting alluvial fan, suggesting that the scarps are much older than the Bonneville high stand. We infer that they were last active in the early part of the late Pleistocene.

The down-to-the-west scarps occur on probable late Pleistocene alluvium in the valley and are locally buried by eolian deposits and modified by modern drainage. These scarps are relatively low and oblique to wind-aligned features but parallel the predominant drainage direction. Field investigations of these scarps were not performed so scarp heights and maximum slope angles are not known. Although Ertec Western, Inc. (1981) indicates Holocene offset, we believe these scarps to be more likely late Pleistocene because they do not appear to be on Holocene deposits and have a subdued appearance on air photos.

Faults along the west flank of the Bald Hills are down to the west (Rowley, 1976; Mackin and Rowley, 1975, 1976). The older scarps are closer to the hills and are highly degraded and discontinuous. They are on only the oldest deposits which are middle to late Pleistocene alluvial-fan deposits from the Bald Hills. A younger, longer, more continuous scarp is found to the west where it is on the toes of these fans (locality G, plate 1) and extends onto younger alluvial-fan deposits. Only the youngest stream deposits are not offset, but modern drainage is disrupted. Alluvial-fan deposits of probable Holocene age are found on top as well as at the base of the scarp. Deposition of fans upstream from the scarp is perhaps caused by back-tilting along short subsidiary faults now buried beneath these fan deposits or by up-to-the-fault warping on the upthrown side. In either case, the disrupted drainage and apparent young age of deposits indicate a probable late Pleistocene age of faulting.

PINTURA AREA

Southwest of Pintura, southeast-flowing Pleistocene drainages heading in the Pine Valley Mountains formed fan complexes in areas of diminished gradient directly west of the Hurricane fault. The fans were partly constructed on Quaternary basalt that was being downfaulted and east-tilted by Quaternary displacements and associated reverse-drag flexing on the Hurricane fault (Hamblin, 1965). The fans are highly dissected, and consist mostly of coarse debris-flow material. Fault scarps with heights as much as 15 m are conspicuous on remnants of the dissected fan surfaces (figure 4). The scarps are mantled with a lag concentration of close-packed cobbles, boulders, and blocks, measuring up to 3 m in diameter. Some

scarps are on the projection of exposed bedrock faults. Most scarps face east toward the Hurricane fault, suggesting antithetic faulting that would be mechanically linked to the Hurricane fault in the same manner as the reverse-drag flexing (Hamblin, 1965). We infer a late Pleistocene age for the scarps consistent with the inferred late Pleistocene deformation on the Hurricane fault. As with the Hurricane fault, the larger scarps near Pintura probably reflect recurrent late Pleistocene faulting.

ENOCH AREA AND WESTERN RED HILLS

Two approximately north-trending west-facing scarps that are on unconsolidated alluvial deposits extend into the incorporated limits of the town of Enoch. The eastern of the two scarps in Enoch has been trenched in several places by local residents in efforts to stimulate flow from springs that issued from the scarp. Layers of brown to dark-brown humic sandy clay are exposed in the trenches east of the escarpment (presumably in the upthrown block). A field sketch was published by Anderson (1980) of the north wall of one of the trenches from which a layer of humic clay yielded a radiocarbon age of 9500 ± 300 yr. The age-dated layer separates indistinctly bedded pebble- and cobble-bearing alluvium below from well-bedded sandy clay above. The coarse alluvium is faulted at the scarp. The overlying well-bedded fine-grained strata show no evidence of thinning or pinching out westward toward the scarp, suggesting that they too may have extended across the scarp prior to its formation. However, excavation of a shallow pit (about 1.5 m deep) at the scarp failed to reveal those beds on the "downthrown" side of the scarp. The maximum scarp slope angles in two profiles are considerably less than what would be expected for a post-9500-year-old scarp (figure 4). Perhaps the maximum slope angle is low because the scarp is on finer grained deposits that might degrade more rapidly than scarps on coarser fluvial and debris-flow gravels; scarps on the latter are represented by the reference curves. If degradation rates are increased, these scarps could be Holocene as suggested by the age data. The ambiguity could probably be resolved by excavating a deep trench on the downthrown block. Because of the uncertainty of the relation of the dated deposits to faulting and the apparent morphologic age of the scarps, we believe that the last surface-faulting event probably occurred during the latest Pleistocene, although Holocene activity cannot be precluded.

Other scarps on alluvium are found west and north of Enoch. Across Cedar Valley 10 km to the west are several scarps at the base of The Three Peaks (Mackin and Rowley, 1976; Rowley and Threet, 1976), and to the north is a northwest-trending series of short scarps beginning 15 km north of Enoch near Highway 130 (north of Rush Lake) and extending to near the edge of the quadrangle just west of the Red Hills (Rowley, 1975). All of these faults cut alluvial deposits of estimated middle to late Pleistocene age except those farthest north, which cut highly dissected alluvium from the Red Hills of probable early to middle Pleistocene age. The youngest scarps are found at the southern end of the group of faults north of Rush Lake, where preliminary profiles and age estimates of the youngest faulted deposits indicate that some of the scarps may be of latest Pleistocene age. Because none of the faults north and east of Enoch were studied in detail in the field, we assign a middle to late Pleistocene age for all scarps except those farthest north which are older and of probable early to middle Pleistocene age.

PAROWAN VALLEY

Scarps on alluvium are widespread in Parowan Valley. The scarps on the west side are on middle to late Pleistocene alluvial-fan deposits and are generally higher and older than those on the east side. The highest scarps on the west side generally face east, and large alluvial fans have been deposited at the base of these scarps. Along major drainages, these post-fault fans have become incised and modern deposition has shifted down fan. Scarps facing uphill (west) are present locally, and major drainages are re-established across these scarps with no areas of blocked or closed drainage remaining. Although all scarps are on relatively steep piedmont slopes, resulting in rapid deposition and re-establishment of drainage, the degree of dissection of faulted deposits and the size of post-faulting alluvial fans suggest a middle to late Pleistocene age for the last surface displacement.

Scarps on the east side of the valley generally face west and are found on lower, nearly flat piedmont alluvial slopes. Some are linear with little erosional dissection, whereas others are detectable only as discontinuous short segments, due to either erosional modification or initial discontinuities in trend and height. The scarps were profiled in two places: (1) NW¼ sec. 19, T. 34 S., R. 9 W. (locality H, plate 1), and (2) W½ sec. 8, T. 33 S., R. 8 W. (locality I, plate 1). A semilog plot of scarp height versus maximum slope angle for these profiles is shown in figure 4. At locality H (plate 1), the fault is exposed in a road cut where the net fault displacement is distributed over a series of faults in a zone about 7 m wide. Though offsets along individual faults in the zone could not be determined, the distribution of net slip over several traces indicates that the morphology of the scarp at this location may not be representative of age. The scarp profile for locality H is plotted on figure 4 as the 5-m-high scarp to the right of other Parowan Valley scarps.

The profiles at locality I (plate 1) reveal a scarp about 2-2.5 m high with maximum slope angles of about 9°, indicating a morphologic age slightly older than 15,000-year-old Bonneville-age scarps (figure 4). The soil on the offset surface has a 25-cm-thick stage I carbonate horizon (Gile and others, 1966; Machette, 1985b) and 15-cm-thick argillic B horizon, suggesting a probable late Pleistocene or early Holocene age for the soil. In general appearance, this scarp is one of the youngest surface-faulting features in the quadrangle and is designated late Pleistocene (Holocene?) on plate 1.

MARKAGUNT PLATEAU

A few highly dissected short scarps on alluvium are in southeastern (Upper) Bear Valley about 25 km north of Panguitch Lake in the Markagunt Plateau (Anderson and Miller, 1979). These scarps and the alluvium on which they are formed are probably middle to late Pleistocene in age, based on a qualitative assessment of scarp morphology and dissection. Vegetation lineaments and concordant eroded fan toes occur between the short scarps, but evidence is insufficient to determine if they indicate connecting fault traces. These Upper Bear Valley alluvial scarps lend credence to our assessment of a middle to late Pleistocene age for activity on range-bounding and bedrock faults in the area. Anderson and others (1987) infer buried range-front faults in the vicinity of the scarps, and also postulate that Quaternary faults cut bedrock in the area.

CENTRAL SEVIER VALLEY NEAR PANGUITCH

A 13-km-long zone of fault scarps on Quaternary alluvium in the area east and northeast of Panguitch, from near Panguitch north to Sand Wash, appears to be the north part of a 25-km-long zone of genetically and temporally related scarps that extend south of Panguitch to near Hatch. Those south of Panguitch include scarps on Quaternary alluvium, pre-Quaternary basalt, and alluvial deposits of the Tertiary Sevier River Formation. They are described in the section on faults in pre-Quaternary bedrock. The scarps in both areas are similar in form, north to northeast trend, and structural setting, and their lack of mapped continuity may be due to erosional destruction by the Sevier River which separates the two areas.

Scarps east and northeast of Panguitch are on isolated remnants of Pleistocene surfaces that are surrounded by and partially buried by a complex of unfaulted Holocene fan and channel deposits. The Pleistocene surfaces are underlain mostly by boulder and cobble-bearing alluvium derived primarily from volcanic rocks in the Sevier Plateau to the east. On the basis of a preliminary study of scarp morphology in the area (discussed as subareas in the next two sections), Bucknam and Anderson (1979) regressed scarp-slope angle on log scarp height to estimate the age of the scarps relative to reference regressions. Subsequent study indicates that few of the scarps represent single surface-faulting events and many, or perhaps all, of those whose anomalously steep mid-slope sections had been interpreted to indicate multiple events may actually reflect stratigraphically controlled contrasts in resistance to erosion. These complications, together with the high degree of scatter in available scarp-profile data (figure 10) preclude meaningful statistical analysis.

A small group of scarps near Sanford Creek in the north part of the area bound a conspicuous horst, whereas those in the south part of the area near Race Hollow are on a Quaternary anticline, the axis of which is marked by a conspicuous keystone graben (Race Hollow), located directly east of the Panguitch airstrip. Those near Sanford Creek contain the clearest evidence of multiple displacements including: (1) fan surfaces showing progressively greater offset with increasing age, and (2) stratigraphic relationships exposed in a trench. The scarps near Sanford Creek are discussed first followed by those near Race Hollow.

Sanford Creek Area

Horst-bounding scarps are on two Pleistocene surfaces of significantly different ages near Sanford Creek. Scarps as much as 12 m high separate offset remnants of a surface that may be as old as middle Pleistocene. Many of these high scarps have a midslope segment that is anomalously steep and appears as a conspicuous dark line on air photos. It is not certain whether the steepness results from a relatively young surface-faulting event or from lithologic control affecting scarp degradation. The trace of the west-bounding fault of the horst trends about N. 25° E. and is defined by a scarp as much as 0.8 m high where it crosses a middle to late Pleistocene surface that is topographically lower than, and obviously younger than, the surface containing the high scarps. The maximum slope angle of this west-facing low scarp is about 3°, and it is formed on a fan surface that slopes 0.5°-1.5° W. As noted by Machette and others (1986), estimating age from scarp slopes that differ from general land-surface slopes by only a few degrees is difficult.

A commercial wood-chip disposal trench about 50 m long, 8 m wide, and 5 m deep, and trending approximately N-S, was excavated across the low scarp on the middle to late Pleistocene surface in October 1987 (locality J, plate 1). Access was granted to the trench for a brief reconnaissance investigation of the fault zone. Faults exposed in the trench strike N. 20° E. to N. 28° E. and are thus intersected by the trench at low angles that are not ideal for graphic representation of offset stratigraphic relationships. Figure 11 is a log of the east wall of the central part of the trench sketched from field photographs. The trenched deposits are typical of those in the distal parts of coalescing alluvial fans in an aggrading alluvial piedmont and are characterized by coarse-grained (sand and gravel) materials. In figure 11, lithologic units are grouped according to their stratigraphic relationship to major unconformities and soils evident in the alluvial-fan sequence. Units are numbered from oldest (1) to youngest (4). Minor cut-and-fill features (channels) within major units are lettered beginning with the

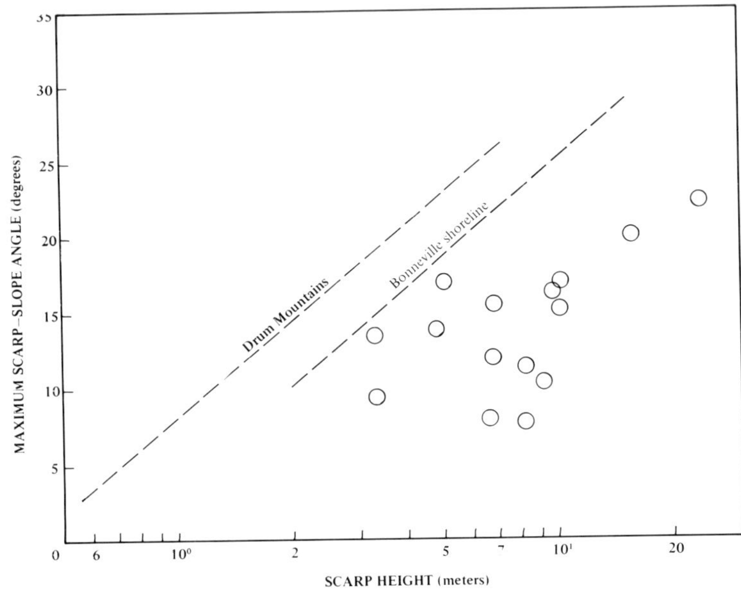

Figure 10. Plot of maximum scarp-slope angle versus log of scarp height for scarps in the Sanford Creek and Race Hollow areas. Labeled dashed lines are reference regression lines on data from scarps of approximately 10,000 yr (Drum Mountains) and 15,000 yr (Bonneville shoreline) from Bucknam and Anderson (1979).

Figure 11. Schematic log of the central part of the east face of a wood-chip disposal trench near Panguitch (location J, plate 1). The log was sketched from a mosaic of field photographs taken at eye height while standing in the bottom of the trench. The trench wall is not vertical (slopes about 75° away from the camera), and the camera angle from the bottom of the trench produces a significant scale inconsistency making unit thicknesses and fault offsets measured from the log less than actual dimensions, particularly near the top. This, and the oblique angle between the strike of faults (N. 25° E.) and the trench wall (north-south) account for the apparent dip of the faults to the south. Faults actually dip 84°-89° to the east and are down to the west. The top of the trench and basal contact of the uppermost unit are approximate because they do not appear on photos used to construct the log. Also, the orientation of the log with respect to the horizontal is approximate and dips of beds and slope of the surface measured from the log are not accurate. See text for discussion of stratigraphic and fault relationships.

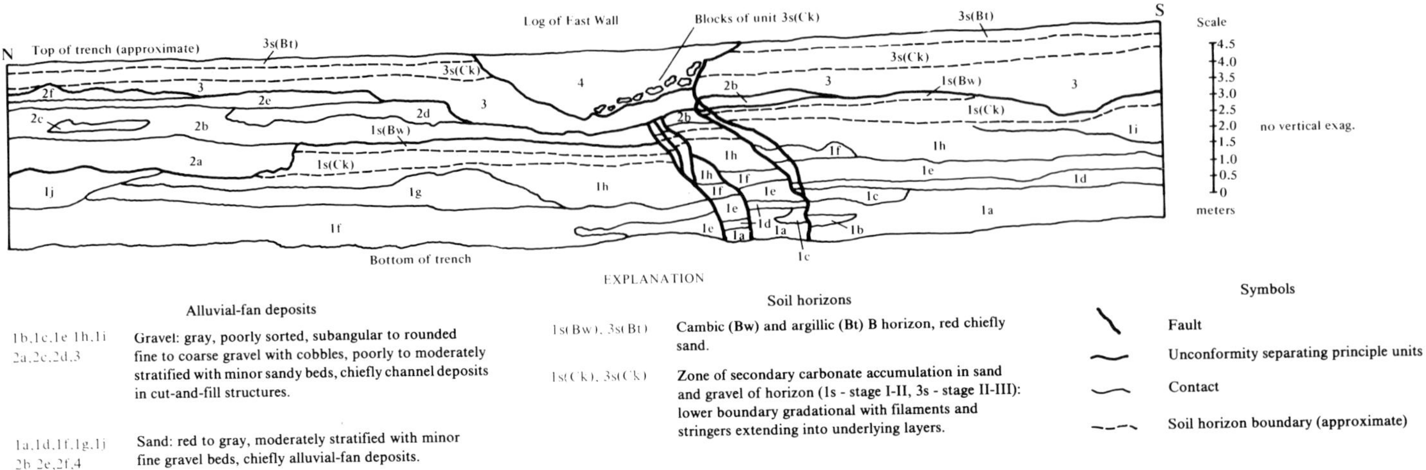

EXPLANATION

Alluvial-fan deposits

1b,1c,1e,1h,1i 2a,2c,2d,3 — Gravel: gray, poorly sorted, subangular to rounded fine to coarse gravel with cobbles, poorly to moderately stratified with minor sandy beds, chiefly channel deposits in cut-and-fill structures.

1a,1d,1f,1g,1j 2b,2e,2f,4 — Sand: red to gray, moderately stratified with minor fine gravel beds, chiefly alluvial-fan deposits.

Soil horizons

1s(Bw), 3s(Bt) — Cambic (Bw) and argillic (Bt) B horizon, red chiefly sand.

1s(Ck), 3s(Ck) — Zone of secondary carbonate accumulation in sand and gravel of horizon (1s - stage I-II, 3s - stage II-III): lower boundary gradational with filaments and stringers extending into underlying layers.

Symbols

Fault

Unconformity separating principle units

Contact

Soil horizon boundary (approximate)

oldest (a). Soils developed on units are designated with an s, followed by a parenthetic horizon designation (Bw, Bt, Ck).

Two faulting events are indicated. They occurred on separate strands of the subvertical fault zone that cuts the nearly horizontal strata. Dip separation during the first event is about 60 cm distributed between two to three western strands of the zone. This deformation is restricted to beds beneath an unconformity at the base of unit 3 (figure 11).

Following the first faulting event, a principal channel shifted into the area, removing much of unit 2 from the upthrown side and depositing a laterally continuous, coarse-grained channel gravel (unit 3). This channel was then abandoned and pedogenesis ensued resulting in development of a soil consisting of a thin A horizon, a reddish, argillic Bt horizon 15-30 cm thick, and an underlying carbonate horizon (Bk and Ck horizons) 55-70 cm thick exhibiting stage II-III carbonate morphology. This soil is better developed than that on unit 1 which predates faulting. In comparing profile characteristics of the unit 3 soil to soils in the sequence described in detail by Machette (1985a) in the Beaver, Utah area, unit 3 appears to be similar to soils dated at 120-140 Ka. The soil is thus probably middle to late Pleistocene (many tens of thousands of years) in age. The soil is on the fan surface that has the 0.8-m-high scarp representing the last surface-faulting event — an event that clearly postdates the middle to late Pleistocene deposit.

The scarp height of 0.8 m agrees very closely with the stratigraphic throw of 85 cm measured across the eastern splay of the fault zone which displaces unit 3 (the parent material for the soil) in the trench (figure 11). That only the youngest of the two faulting events seen in the trench is expressed by the surface scarp is consistent with evidence of strong erosional planation, especially on the upthrown block, prior to deposition of unit 3 and subsequent to the first faulting event (figure 11).

The age of the last surface-faulting event is not well constrained because it is not clear if unit 4 or the soil on unit 3 is faulted. In the trench, the scarp had been removed by grading and thus its relation to units 3 and 4 is unknown. A steep east contact between unit 3 and the lower part of unit 4 can be interpreted either as a fault-generated free face or an undercut channel margin. If it is a fault-generated free face, the faulting postdates principal calcic soil development and cutting of the channel. If it is an undercut channel margin, the channel cutting postdates principal calcic soil development, but the temporal relation between faulting and soil development is unknown. Either interpretation would account for the colluvial wedge containing caliche-cemented gravel which extends westward from the steep contact to the base of the channel in unit 4 (figure 11). Except for this basal colluvium along the east bank, the channel is filled with homogeneous coarse sand representing generally lower energy channel fill, perhaps derived more from slopewash and bank erosion than from channel deposits. A very weakly developed zone of disseminated carbonate is present in unit 4, and soil development appears roughly similar to that described in Holocene soils by Machette (1985a) in the Beaver, Utah area. This soil and the parent materials in which it formed probably postdate the last faulting event which, accordingly, is considered to be late Pleistocene.

Because of uncertainties in the relation between soil development and the last faulting event, little can be concluded regarding the timing of the event and the time between events. However, the preservation and linearity of the small (80 cm) scarp in the middle to late Pleistocene alluvial surface suggests a late Pleistocene age for the last event. Anderson and Rowley (1987) report that these faults cut Holocene alluvium in two areas where vegetation lineaments are visible in Holocene alluvium on trend with the fault. There are no scarps or offset beds associated with these lineaments, and we believe they are probably a result of ground water concentrated along the late Pleistocene faults buried beneath the unfaulted, shallow Holocene alluvium.

Race Hollow Area

Pleistocene fan surfaces in the Race Hollow area exhibit opposed structural rotations as much as a few degrees away from a central graben trending N. 20° E. occupied by Race Hollow. The tilted surfaces define a gentle faulted anticline in the hanging wall of the Sevier fault. Pleistocene surfaces on basalt and pediment gravels in the area south of Panguitch reveal a faulted anticline of similar trend and Wagner (1984) reports a northeast-trending gentle syncline between Panguitch and Panguitch Lake, indicating that the fold in the Race Hollow area is not unique. Displacement directions on the faults are generally down toward the anticline axis thereby counteracting the structural relief due to folding. In general, the positive structural relief resulting from the gentle, open fold exceeds the negative relief attributable to faulting indicating that folding is an important aspect of Quaternary deformation in the Panguitch area. Scarps in the Race Hollow area are as much as 27 m in height and have maximum slope angles that range from 8° to 23° (figure 10). The oldest faulted surfaces are probably as old as middle Pleistocene. As with the Sanford Creek area, the scarps and tilted surfaces are bisected by several west-flowing streams that drain from the Sevier Plateau into the Sevier River. Each exhibits erosional downcutting, the early stages of which may have coincided with folding and faulting. Mature calcic soils representing many tens of thousands to more than 100,000 years of pedogenesis are on the erosionally modified landscape, and in many places these soils are buried by Holocene fluvial deposits. Here, radiocarbon ages reported by Anderson (1980) indicate that burial at one locality occurred between 4000 and 5000 years ago.

About 4 km north of the Bryce Canyon turnoff (intersection of Utah Highways 89 and 12), a low-lying, evenly graded surface, possibly equivalent in age to the trenched surface at Sanford Creek, is displaced vertically less than 1 m. This is the only evidence we are aware of that late Pleistocene surface faulting as young as the youngest event recorded in the trench (locality J, plate 1) may have occurred in the Race Hollow area. We infer that the last surface faulting in the area occurred during middle to late Pleistocene.

SEVIER PLATEAU

In the Sevier Plateau, an east-west-trending scarp on alluvium parallels Hunt Creek in Johns Valley 27 km east of Panguitch and is roughly concordant with an east-west-trending bedrock fault in the area which defines the south flank of a large Pleistocene landslide 3 km to the west. Although the alluvial scarp parallels a group of stream-terrace scarps, it is thought to represent a fault scarp rather than a stream-terrace scarp. The surficial characteristics of the alluvium above and below the scarp are similar, indicating a displaced surface rather than two separate stream terrace surfaces of different ages. Also, a short parallel scarp of lesser height occurs near the west end of the longer scarp, and it dies out both east and west in a manner more characteristic of fault scarps than stream-terrace scarps. The maximum height of the longer scarp is about 1 m, and the degree of degradation of the scarp indicates a probable late Pleistocene age. The faulted deposits are mapped by Rowley and others (1987) as Pliocene and Pleistocene in age.

Many other parallel scarps are present in the area along Hunt Creek, and we consider it possible but unlikely that any of the others are fault-related. The mapping by Rowley and others (1987) does not identify any Quaternary faults in the area, and it is possible that none of the scarps are fault-related. However, for the reasons given above, we believe that at least the scarps shown should be considered to represent faults, pending more detailed investigations.

SCARPS ON QUATERNARY BASALT

Our current understanding of long-term displacement rates on the several major faults described above is derived largely from offsets of age-dated or age-estimated basalts. Parts of those faults at or near offset basalts are discriminated separately on the map (plate 1). In addition to main-fault displacements, the map shows several clusters of lesser faults that displace Quaternary basalts. These are described separately here.

VOLCANO MOUNTAIN NEAR HURRICANE

Hamblin (1970b) devised a scheme for assigning relative ages to basaltic lavas in southwestern Utah and adjacent Arizona on the basis of the extent to which they have been modified by erosion. Coupled with radiometric dating of some of those flows, it is possible to differentiate between two to three Quaternary-age groups. W.K. Hamblin (written communication, 1987) utilized this scheme in mapping the St. George and Hurricane 15-minute quadrangles where he separated Quaternary from Tertiary basalts and subdivided Quaternary flows into two map units. Faults in Pleistocene basalt at Volcano Mountain directly southwest of Hurricane are taken from unpublished mapping of Hamblin. The age of last surface displacement cannot be constrained more closely than Pleistocene.

NORTH HILLS AND CROSS HOLLOW
HILLS NEAR CEDAR CITY

Faults in the North Hills displace basalt down toward the axis of a late Cenozoic uplift. The uplift trends north and is defined by a series of outward-tilted fault blocks. The dip of the basalt (age dated about 1 Ma) is conspicuously less than dips in older rocks (Anderson and Mehnert, 1979). On the east the basalt dips 10°-15° east toward the Hurricane fault and on the west the basalt dips westward as much as 30° (Anderson and Mehnert, 1979). Because the basalt was probably near

horizontal when extruded, most or all of the current physiographic expression of the North Hills results from post-1-Ma deformation, though the dip of the older rocks indicates there was pre-1-Ma deformation as well. The Quaternary part of this faulted uplift has a form resembling the faulted Quaternary folds in the Panguitch area. We assign a middle to late Pleistocene age to the last faulting event because it is unreasonable to assume that all deformation ceased prior to about 750 ka.

High-angle normal faults in the Cross Hollow Hills cut basalt with a reported age (Anderson and Rowley, 1975) similar to that in the North Hills. Two of these faults are exposed in highway excavations along Interstate Highway 15 in the topographically low area between the Cross Hollow and North Hills. As part of a dam-safety study, faults at the southeast margin of the Cross Hollow Hills were trenched (Earth Science Associates, 1982), but nothing more definitive was found than verification that the basalt is offset. Our map shows the location of several short scarps that represent displacements in basalt estimated to be less than 10 m. Whether faults in the Cross Hollow Hills have been active in the late Pleistocene is not known. The age of last surface displacement is estimated to be early to middle Pleistocene because we lack specific age data on the faulted basalt.

ENOCH GRABEN

A basalt flow that yielded an age by the K-Ar method of 1.28 ± 0.4 Ma (Anderson and Mehnert, 1979) is part of a series of Quaternary flows that are offset by several north-striking normal faults 5 km north of Enoch. The faults define a narrow horst in basalt along the west boundary of the Enoch graben— a feature that is expressed by a combination of faults in alluvium (discussed previously), faults in basalt, faults in bedrock, and aligned cinder cones that surmount the lavas along the axis of the graben. Throw of basalt on individual faults ranges from 0 to 50 m, but the west-bounding graben fault may have a Quaternary throw of more than 50 m. Because some of the faults are expressed as scarps on alluvium and are inferred to have had late Pleistocene surface offset with good examples in the town of Enoch, we infer that displacements of the basalts also occurred in the late Pleistocene, and therefore we distribute the 50 m of throw over the past 1.3 m.y.

NORTHWESTERN KANE COUNTY

Northeast-striking faults cut Quaternary-age basalts in the Bear Creek area of northwestern Kane County. Best and others (1980) report a K-Ar age of 0.8 Ma on flows in the southern part of the area, with surrounding flows ranging in age from 0.36 to 1.4 Ma. Detailed maps of individual flows have not been made so the correlation of the age-dated units with faulted units is not certain. Faulting is widespread in the basalts and photogeologic studies indicate that the uppermost flows are displaced in many places but that scarp-slope angles are subdued. Cashion (1961) indicates 15 m of throw in the basalt. The last displacement is inferred to be middle to late Pleistocene in age because of subdued morphology. These faults extend beyond the basalts into pre-Quaternary bedrock where their traces are marked by surface morphology similar to that seen on the basalt, and a similar age of last displacement is inferred (plate 1).

FAULTS IN PRE-QUATERNARY BEDROCK

Geomorphic criteria used for assessing or inferring the age of last surface faulting on major faults such as the Washington, Hurricane, and Sevier described in foregoing sections are also used for smaller unnamed faults in pre-Quaternary bedrock, most of which are found on the Markagunt Plateau (plate 1).

MARKAGUNT PLATEAU

Photogeologic inspection of the Markagunt Plateau reveals a conspicuous northeast-trending physiographic grain composed of steep-sided sharp-crested ridges, narrow valleys, scarps, drainage alignments, elongated closed basins, hillside trenches or depressions, and volcanic-cone alignments. The grain is least conspicuous or absent in areas covered by Quaternary landslides or basalt flows. Some bedrock scarps terminate at or are interrupted by cross-strike landslides, indicating that those scarps are older than the landslides. Parts of other bedrock scarps are along landslide boundaries, suggesting that landslides commonly exploit fault-weakened rock.

Scarps are especially concentrated in the northwest Markagunt Plateau (plate 1) where the general physiography suggests vigorous erosion of an unstable terrain. Field studies of these scarps generally were not made. The apparent age of the scarps as determined from air photos varies greatly, and some are certainly latest Pleistocene. Though many of the scarps mark the surface trace of large-displacement bedrock faults, it is unclear to what extent the surface instability reflects youthful seismogenic faulting versus gravitational spreading and (or) collapse. Some of the mapped scarps may be the lateral margins of landslides. The scarps are most common in the part of the Markagunt Plateau where Tertiary and older sedimentary rocks are overlain by layers of resistant welded ash-flow tuffs (Rowley and others, 1978), which is a common environment for landslides and gravity spreading features elsewhere on the Colorado Plateau (R.F. Madole, oral communication, 1988). Because of the varied apparent age and the uncertain association to seismicity, the scarps of the northwest Markagunt Plateau are all grouped into the middle to late Pleistocene category of last surface displacement. Determination of whether or not this is an area of concentrated Pleistocene tectonic deformation must await detailed studies.

HILLS BETWEEN PANGUITCH AND HATCH

The hills west of the Sevier River between Panguitch and Hatch are part of the erosionally dissected east flank of the east-tilted Markagunt Plateau block (Anderson and others, 1975). They consist mostly of Sevier River Formation of late Tertiary age overlain by basalt that yielded a K-Ar age of 5.3 ± 0.5 Ma (Harald Mehnert, written communication, 1988). Structurally, the hills appear to be the southerly continuation of the Quaternary faulted anticline in the Race Hollow area. An eastern anticline and a western syncline (plate 1), each with limbs that have average dips of about 5°, are conspicuous in the basalt. These structures are east of and parallel to a syncline in the Panguitch Lake area reported by Wagner (1984). Most fault scarps are on the basalt, some are on the Sevier River Formation, and one is on post-basalt alluvium of presumed Quaternary age. The scarps strike predominantly north-northeast or northwest, and intersect at an axial keystone graben where they appear to be coeval. Except for the

keystone-graben scarps, which are as much as 25 m high, most scarps are only a few meters high. Though small, the scarps were probably not formed in a single event because they appear to be genetically and kinematically related to the folds which probably grew over a long period. Here, as in the Race Hollow area to the north, the total deformation resulting from folding far exceeds that due to faulting. We speculate that without the component of uplift associated with the folding, much of the area between Panguitch and Hatch would be part of the Sevier River Valley and blanketed by Quaternary alluvium.

Though we place the last faulting event in the broad age range of middle to late Pleistocene, we strongly suspect that deformation continued into the latest Pleistocene and possibly into the Holocene. Some drainages are disrupted, as indicated by the presence of several closed basins in graben in the area. Where the Sevier River crosses the northward projection of the anticline axis, its channel pattern is suggestive of a disequilibrium condition characteristic of active uplift (David Jorgensen, oral communication, 1988). This suggests a component of Holocene uplift.

The horst in the Sanford Creek area and the faulted folds northeast and south of Panguitch are subparallel to the Sevier fault, and we suspect that they are genetically related to that fault. The clusters of closely spaced scarps probably represent faulting with nominal depth penetration and, therefore, nominal potential for hazardous seismicity. The folds are the predominant structures, and we suspect that they have formed aseismically. If the inferred genetic tie to the Sevier fault is correct, the chief importance of the faulted folds from the standpoint of seismic hazards is that they suggest significant middle to late Pleistocene deformation on that large fault.

QUATERNARY STRUCTURES INFERRED TO BE NONSEISMOGENIC

Several Quaternary features, including the most conspicuous indications of Holocene surface disruption and volcanism in the quadrangle, are interpreted to be nonseismogenic, or at least incapable of generating large surface-faulting earthquakes. Though it may appear unorthodox to devote considerable space to describing probable non-seismogenic features, we believe that their preclusion from seismic hazards evaluations is important for avoiding an overestimate of the seismic potential.

DEFORMED GRAVELS BETWEEN THE WASHINGTON AND HARRISBURG DOMES

The Washington and Harrisburg Domes, located southwest of Hurricane, are two of several northeast-trending structural highs along the axis of the Virgin anticline in the St. George-Washington area. The Virgin River flows west between the two domes. In an area of about 1 km² between the Virgin River and the exposed bedrock forming the northeast end of the Washington Dome (locality K, plate 1), a thick sequence of gravels and sands is faulted and tilted to dips averaging about 20° but as much as 80° near the faults. Igneous-rock clasts in the gravels are exclusively basalt and latite porphyry. Whole-rock K-Ar age determinations of two basalt clasts are 1.3 ± 0.2 and 1.6 ± 0.4 Ma (Harald Mehnert, written communication 1988), indicating that these clasts are derived from terrain containing early Pleistocene basalt flows. The Pine Valley Mountains 15-20 km to the north represent the only potential source of this two-type igneous-clast assemblage. The attitudes of bedding in the gravels suggest a southwest-facing nose of a fold, but dip directions are variable. Subsurface projection of the gravel attitudes suggests that the beds dip into the oppositely dipping Permian and Triassic rocks that form the northeast nose of the Washington Dome. The structural relationships require some kind of accommodating structure separating the oppositely dipping strata. The gravels are many tens of meters thick, though there may be some repetition due to faulting. There is at least one and probably two angular unconformities in the gravels suggesting syndepositional deformation. Because we do not recognize similar gravels elsewhere in the area, we believe that they are restricted to a highly localized depocenter resulting from subsidence accompanying dissolution of subsurface evaporites whose presence can be predicted in the structural low between the domes.

SCARPS IN THE HURRICANE CLIFFS AT BRAFFITS CREEK

In the Cedar City-Parowan monocline, the complex assemblage of locally derived debris-flow and landslide deposits records major deformation of probable late Tertiary age. These stratigraphically and structurally complex deposits are further complicated by a second generation of currently active landsliding and by numerous faults, some with conspicuous scarps. Nowhere in the monocline are the details of the com-

plexities better exposed than along the West Fork of Braffits Creek where a spectacular episode of modern erosional downcutting has formed an inner gorge (figure 12) with about 5 km of nearly continuous exposure. Physiographic evidence for youthful deformation as well as cross-sectional sketches of the Braffits Creek area were reported by Anderson and Bucknam (1979b) and radiometric evidence for Holocene faulting was reported by Anderson (1980). Fault-slip and geodetic studies conducted since those reports highlight the importance of lateral motions in shaping the complex structural picture in the Braffits Creek area. Inner-gorge exposures reveal many faults, some within a single lithologic unit and others juxtaposing units of highly contrasting lithology. Many faults have well-developed subhorizontal striations. Unfortunately, the faulted units are not part of an orderly succession with good-quality marker beds allowing for the determination of displacement magnitudes. Gouge zones are common and are as much as 50 cm wide, suggesting displacements in the range of tens to hundreds of meters on some of the strike-slip faults.

Of possible special significance are fault-slip data from an exposure where ^{14}C age data indicate late Holocene faulting (figure 13). The main high-angle fault separating the flat-lying Holocene alluvium from the faulted and tilted older sediments was previously interpreted as a reverse fault (Anderson, 1980). Although slip lineations were not found on this main fault, they were found on a minor splay fault and on two small faults within 1 m east of the main fault (marked G and F in figure 13). All of the slip data indicate predominately strike-slip motion. Also, striae indicating predominately strike-slip motion were

Figure 12.   *Cross sectional sketch showing reconstructed paleosurface (dashed line) that may have existed at the time a fence was constructed around 1920 in contrast to existing grade (heavy solid line) across part of inner gorge of Braffits Creek. The restored wires and posts are still connected and currently rest on the existing grade. It is reasonable to estimate 20 m of erosional downcutting in 60-70 years. This rapid erosion has stimulated numerous small landslides and slumps along the flanks of the gorge.*

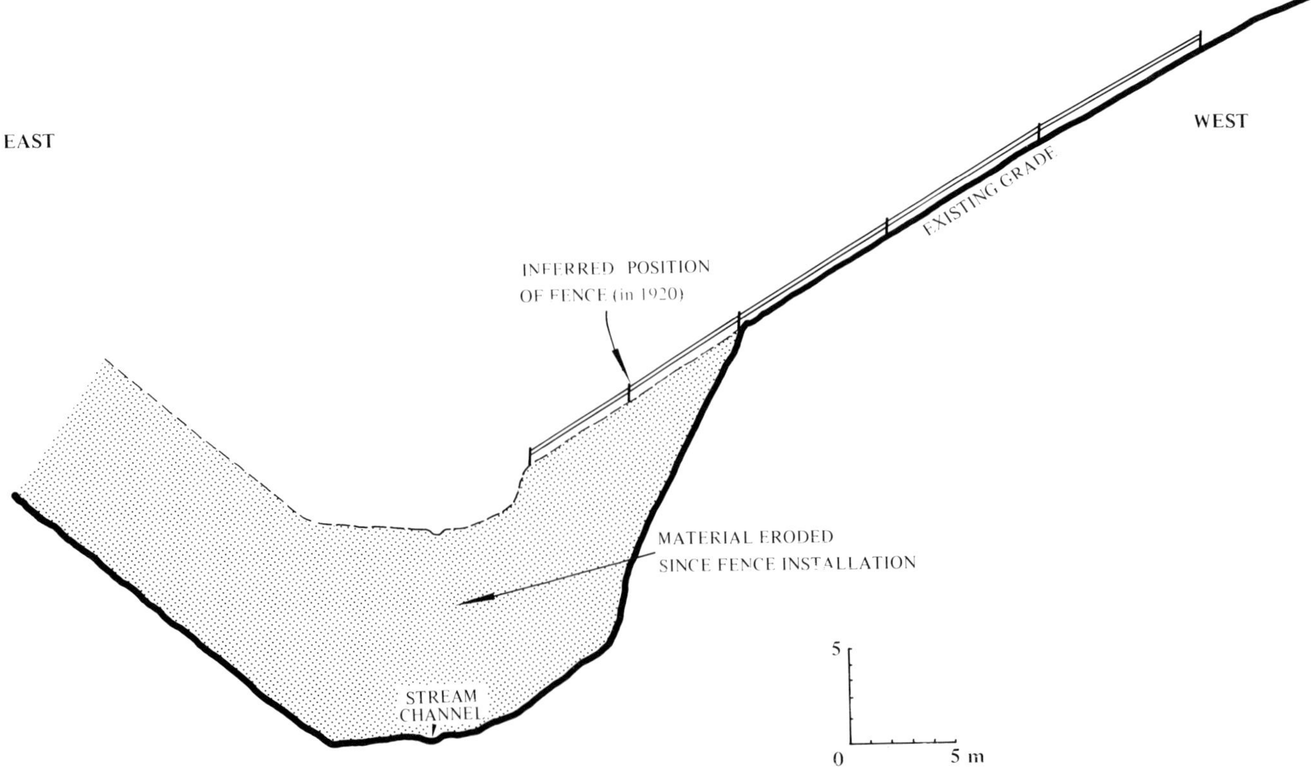

found on three faults that parallel the main fault and have significant apparent displacement (B, C, and D, figure 13). A component of dextral slip approximately colinear with that of faults B, C, and D is inferred for the main fault based on the sense of stratal drag of beds in the hanging wall. The inferred slip sense is thus normal dextral which is consistent with the absence of compression-related minor structures that would be expected in the mechanically weak unconsolidated Holocene alluvium west of the main fault if main-fault displacement

sense were reverse as reported earlier. Normal-dextral displacement on the main fault introduces the possibility that the Holocene channel-fill deposits in the hanging wall are correlative with those of the footwall, as indicated by the similar ^{14}C ages from these deposits (figure 13), rather than being deposited at different times separated by an episode of faulting, erosion, and deposition as inferred earlier (Anderson, 1980). The earlier interpretation was made prior to measurement of slip lines on the faults and is in conflict with the age data.

Figure 13. *Sketch looking south-southeast at exposure along West Fork Braffits Creek about 2.5 km south of confluence with East Fork. The internally faulted sediments consist of well-bedded pale-red silt, sand, and gravel that have an average attitude of N. 20° E., 55° SE. (trace of bedding indicated by form lines). These sediments consist of a relatively homogeneous clast assemblage derived from lower Tertiary sedimentary rocks. They are sheared and brecciated and are in high-angle contact (fault labeled A) with flat-lying heterogeneous alluvium to the west of man and are channeled and filled by coarse alluvial debris to the east of man. Large labelled dots are sample sites for which radiocarbon ages are shown at end of leaders (Anderson, 1980). Faults labeled B, C, and D are conspicuous shears with orientations similar to that of fault A as shown in the inset lower-hemisphere projection. The shafts of the labeled arrows represent strial in the respective faults. No striations were found on fault A. Striations on small-displacement microfaults directly east of fault A (E, F, and G shown by dashed lines and labeled arrows in inset) indicate a major component of strike slip. Together the age and fault-slip data support the inference of important Holocene dextral slip on fault A.*

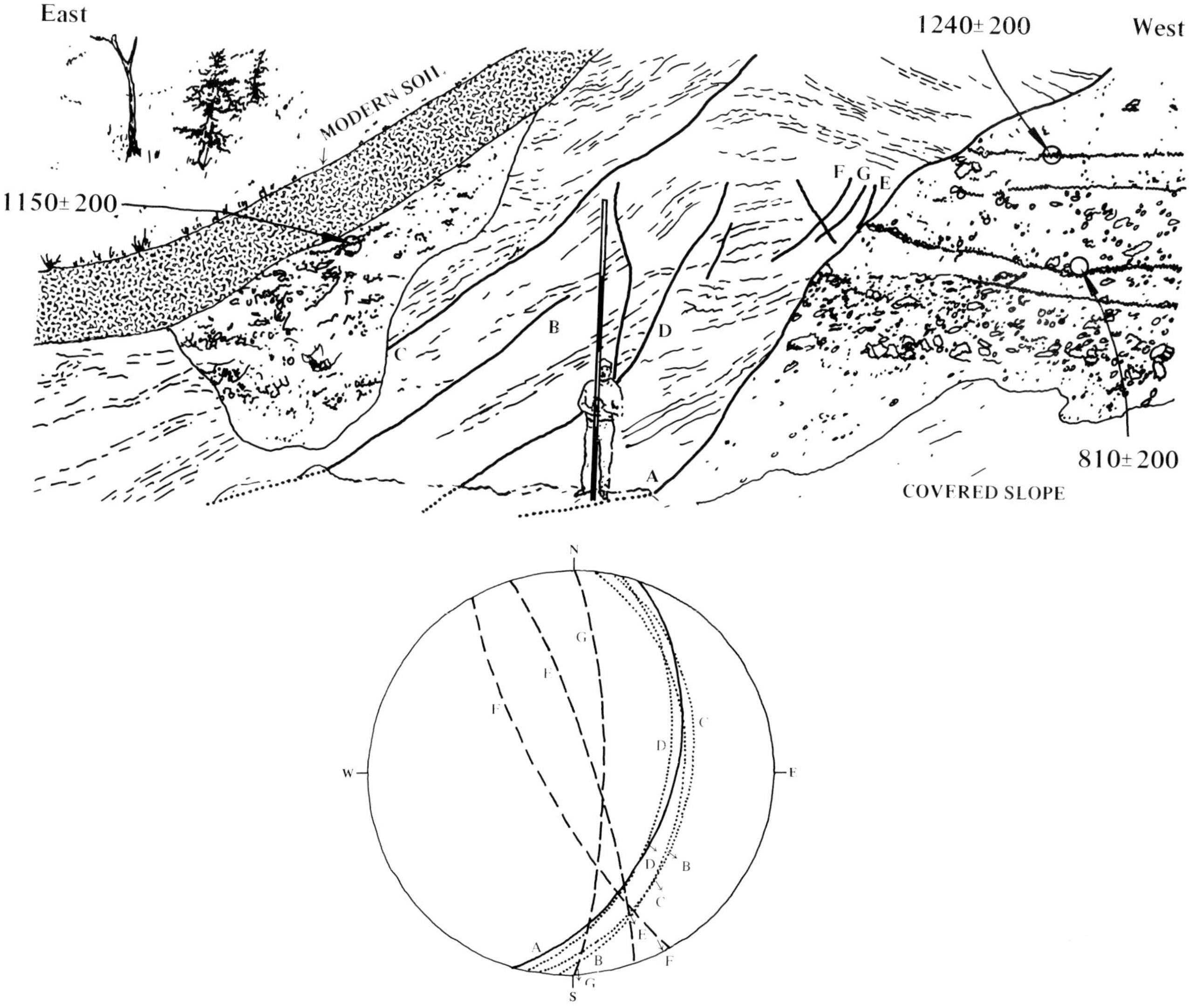

In October 1977, a geodetic network was installed across Braffits Creek between the two localities marked L (plate 1) for the purpose of monitoring the apparently large young strain indicated by landforms in the area. The network consists of an eight-station set of bridged quadrilaterals. It is about 1 X 4 km with the long dimension oriented northwest (figure 14). Stations 1 and 2 of the network are on alluvial fans 400 and 600 m, respectively, northwest of the bedrock base of the Hurricane Cliffs. They constitute a basin-range base line to which any position shifts of other stations in the network can be compared. Station 7 is located on Cretaceous sedimentary rocks where good exposures preclude significant fault or fold deformation over an area about 1 X 3 km. Any position shift of station 7 relative to the base line should represent movement of a large coherent block with a thickness of at least 1 km. Stations 3 through 6 are located at topographic high points on Cenozoic volcanic rocks interpreted to represent separate bedrock blocks. These bedrock blocks are part of the intensely deformed terrane referred to above and are probably separated from one another by faults or fault systems. Station 8 is located on a slightly deformed part of the same structural block of Cretaceous sedimentary rocks that contains station 7.

tectonic deformation is indicated. Lateral shifts of this magnitude are consistent with widespread evidence of strike-slip faulting in the Braffits Creek area and with the inferred large-magnitude late Holocene dextral slip on at least one fault. A displacement rate of 10 mm/yr similar to the maximum geodetic signature is reasonable for the Holocene fault.

Recognition of lateral displacements in the Cedar City-Parowan monocline is a complicating rather than clarifying factor. No single deformational model seems capable of integrating all aspects of the neotectonic deformation. Horizontal displacements are recognized in the earthquake and fault-slip records of the central Sevier Valley area, Utah (Anderson and Barnhard, 1987; Arabasz and Julander, 1986), where locally they dominate those records. There, also, the horizontal displacements complicate rather than clarify the neotectonic picture.

The high deformation rates in the Braffits Creek area indicated by surface disruption (Anderson and Bucknam, 1979b), faulted late Holocene deposits, and geodetic measurements are apparently unaccompanied by seismicity above a background threshold of about M_L 3.0. Though critical data are lacking, landforms throughout much of the Markagunt Pla-

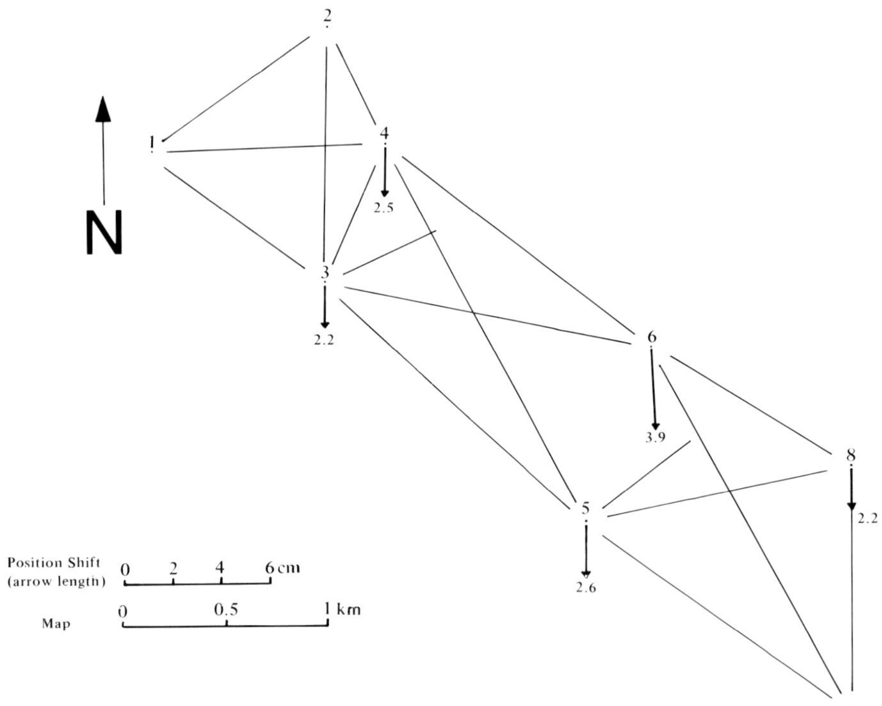

Figure 14. Diagram showing horizontal distribution of eight-station trilateration network established in 1977 in the Hurricane Cliffs across Braffits Creek between the two localities marked L (plate 1). Light-weight arrows and adjacent numbers give azimuth and magnitude (arrow length in cm) of position shifts of stations 3 through 8 relative to stations 1 and 2 between October 1977 and October 1981. Note that station 7 showed no shift relative to stations 1 and 2, suggesting that a corridor across a strain field is captured within the network.

The eight-station network was re-surveyed from October 5-9, 1981, approximately 4 years after it was established. Significant changes in relative horizontal and vertical position of the stations occurred between 1977 and 1981 assuming stations 1 and 2 are fixed to avoid rotational error from being introduced into the 1981 adjustments. Station 6 shows the largest position shift of 39.15 mm. The relative position shifts of stations 3 through 8 are shown graphically in figure 14 where the relative shifts, which are all southerly, are greatly magnified on a planimetric plot of the network. Because the position shifts are in a direction opposite to the topographic gradient, gravity sliding is precluded as an explanation and

teau suggest that high rates of relatively aseismic deformation may be widespread. Graben and closed basins along the Hurricane Cliffs may reflect tensional collapse of the flanks of a rising and spreading monocline or simply spreading in the margins of the rapidly rising Markagunt Plateau block. If so, most of this deformation is probably a superficial (thin-skinned) extensional response to major block uplift and would appear aseismic at fairly low thresholds. If, on the basis of geodetic data, the strike-slip faulting is to be considered tectonic in nature, its lack of accompanying seismicity suggests that it too is probably thin-skinned — perhaps a shallow, brittle, response to lateral ductile flow.

SCARPS ALONG THE SEVIER FAULT
NEAR ALTON

North and northeast of Alton the trace of the Sevier fault is a southeast-facing obsequent fault-line scarp (locality M, plate 1). This unique physiography results from the Cretaceous rocks in the upthrown block being less resistant to erosion than the Tertiary and Quaternary rocks of the downthrown (western) block. In fact, fault-front deposits of colluvium and conglomerate mapped by Doelling and Davis (1989) as Quaternary alluvial gravels are unusually well indurated (possibly as a result of ancient mineralizing ground-water discharges along the Sevier fault) making them more resistant to erosion than the underlying Tertiary strata. These Quaternary deposits commonly form a conspicuous caprock (figure 15) and have probably contributed significantly to the reversal of topography along the fault by retarding erosional downcutting of the downthrown block.

Figure 15. *View uphill and northwest from point marked by A in figure 16B showing highly indurated resistant coarse fault-front colluvium of probable Pleistocene age. The colluvium is on the downthrown side of the Sevier fault but is well exposed here, owing to the development of an obsequent fault-line scarp. The photographer is standing on the main fault trace.*

Within the uneven topography of the obsequent scarp, a very conspicuous west- (uphill-) facing scarp 2 km long and as much as 4.8 m high is found on the main trace of the Sevier

fault. The scarp is formed on well-lithified conglomerate or breccia containing angular clasts of yellowish-gray and pinkish-gray sandstone. The origin and age of the conglomerate are not known. Where the scarp is more than about 2 m high, its lower part is formed on greenish-brown silty shale or medium-grained sandstone of probable Cretaceous age. In profile, the scarp is conspicuously compound with the lower part steepest (figure 16A). At one locality along the scarp, the uphill-side-down displacement sense has produced a small sediment trap (figure 16B). Scarps on transverse ridges to the south of the sediment trap (figure 16A and 16C) are between 4 and 5 m high, and we infer this to be the maximum depth of closure of the trap. Currently, the trap is not filled with sediment and has about 1 m of topographic closure over an area of about 1.6×10^3 m². Uphill-side-down scarps extend away from the trap along the fault trace and serve to funnel all erosional detritus, including debris-flow deposits, from an area of about 40×10^3 m² into the small trap. Because it is not filled with sediment, we estimate that the trap and the scarp bounding it are only a few thousand years old.

The possibility exists that there is little or no tectonic throw associated with formation of the uphill-facing scarp. If straight-line segments of the crestlines of ridges that are transverse to and cut by the scarp (well-illustrated in figure 16C) are projected across the scarp they show approximate concordancy. If it is assumed that the crestline had relatively constant slopes prior to scarp formation, the scarp may simply be the east boundary of a narrow zone of asymmetric subsidence aligned along the main trace of the Sevier fault. Though crestline projections cannot be made so accurately as to preclude tectonic throw, we assume that little, if any, of the approximately 5-m-high scarp reflects tectonic throw. An important implication to this assumption is that the scarp probably does not reflect earthquake-related surface faulting. Regardless, it would be highly anomalous to restrict 5 m of seismogenic surface faulting to a 2-km-long segment of a through-going major block-bounding fault presumed to extend to depths of 15 km or so (Arabasz and Julander, 1986; Anderson and Barnhard, 1987).

Instead, we suspect that the scarp was caused by gravity-driven yielding of mechanically weak Cretaceous rocks east of the fault. The rapid erosional downcutting that produced the obsequent main scarp created a fault-parallel ridge with the trace of the Sevier fault in the eastern ridge flank. Uphill-facing scarps caused by gravity-driven yielding of oversteepened ridges are common (Beck, 1968; Radbruch-Hall and others, 1977) and have a sound mechanical basis (W.Z. Savage, written communication, 1988). Landslides are widespread in the nearly flat-lying Cretaceous rocks east of the fault (Doelling and Davis, 1989), attesting to the susceptibility of those rocks to gravity-driven mass movement. We suspect that the damage zone of the Sevier fault served as a steep breakaway boundary for minor lateral bedding-parallel shifting of a rock mass less than 100 m thick, and that the scarp results from asymmetric collapse associated with the lateral shift. That gravitational stresses and pore pressures could have existed at levels high enough to satisfy bedding-parallel yield conditions in the Cretaceous rocks is, of course, inferred. A similar linear trough parallel to the Sevier fault occurs in less steep landslide terrain east of Black Mountain about 15 km to the south. This trough is less well-defined and more obviously the result of landsliding, but it provides further evidence for gravity-driven mechanisms in the area.

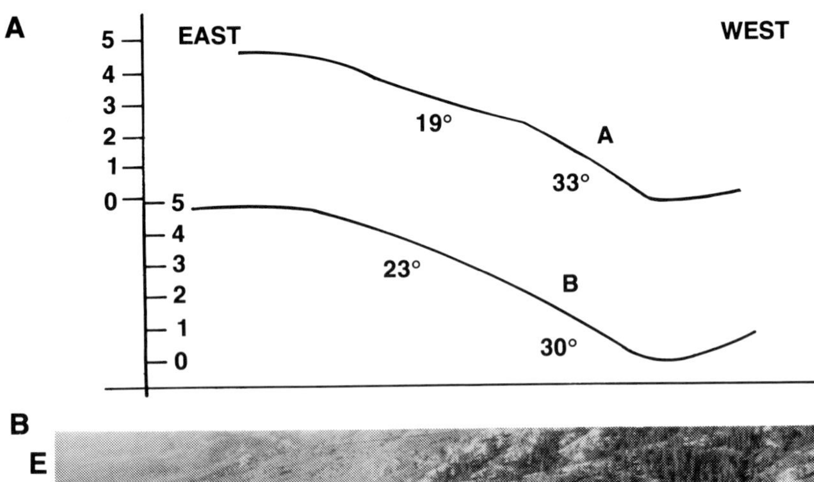

Figure 16. A) Profile sketches of uphill-facing (west) scarps along ridges cut by the main trace of the Sevier fault. The ridges are identified as A and B in figure 16C, corresponding to the profile labels; vertical and horizontal scales (meters) are equal. B) View south-southwest along the surface trace of the Sevier fault north of Alton showing the irregular topography characteristic of the east-facing obsequent fault-line scarp in this area. Photographer is standing on the main trace. The line of tree shadows cast onto the floor of a small closed depression (sediment trap) in the center of the photo conspicuously marks the trace of the uphill-side-down scarp. The photos in figures 14 and 16C are taken from point marked by A. The cliff face in figure 15 is seen upslope (west) from the closed depression. C) Looking south-southwest along trace of Sevier fault from point marked by A in figure 16B showing profiles of back-lighted ridge crests cut by fault. Note that the uphill-facing scarps at A and B mark the east boundary of low parts of the ridge crests across which the ridge crests can be projected as apparent prefaulting continuous lines.

BASALTIC VOLCANIC VENTS

Rocks younger than 5 Ma of primarily basaltic composition are exposed within a northeast-trending zone referred to as the St. George zone by Smith and Luedke (1984). Many flows in the zone are surmounted by cinder cones. On the basis of cone morphology, most are certainly Quaternary but some may be as old as latest Tertiary. Plate 1 shows the location of known or inferred Quaternary-age basaltic vents based on published compilations (Luedke and Smith, 1978; Hamblin, 1970b; Hintze, 1980), unpublished mapping in the St. George and Hurricane quadrangles by W.K. Hamblin, and photogeologic study by the writers. The choice to include a comprehensive compilation of Quaternary vents on the map is guided by the suggestion of Koons (1945) that alignments of basaltic vents of Quaternary age in Arizona are structurally controlled.

Whole-rock analyses of major elements (Best and Brimhall, 1974) and trace-element patterns (Fitton and James, 1986) of latest Cenozoic mafic lavas from the transition zone in southwestern Utah and adjacent Arizona indicate that the magmas were derived from lithospheric mantle. The role of crystal fractionation during magma ascent or during residence in crustal chambers in producing observed chemical variations is thought to be small (Best and Brimhall, 1974). Chemical variations are inferred to have resulted mainly from partial melting in the lithospheric mantle. On the basis of these petrogenetic interpretations, it is reasonable to assume that eruptions originated at great depths and that magmas ascended through the entire crust. Preferential orientations of dike swarms or alignments of volcanic centers should, therefore, reflect deeply penetrating fracture systems that served as channelways for magma escape.

There are eight alignments of Quaternary volcanic vents in the Cedar City quadrangle that can be confidently identified. They are enclosed within whole brackets near their terminations on plate 1. The alignments include from 3 to 28 vents and range in length from 2.5 to 11.5 km. Additional vents are positioned at widely spaced intervals on or approximately on the projection of three of those alignments (marked by incomplete bracket on plate 1). Inclusion of the less-confident members raises the number enclosed by a single alignment to 32 vents and the maximum alignment length to 37 km.

The shortest alignment is the three-cone group along the axis of the northeast-trending Enoch graben. Because this graben is formed in locally derived Quaternary basalt, a high degree of confidence is attached to the structural significance of the axially positioned short cone alignment.

The most conspicuous alignment of Quaternary basaltic vents is the 28-member group at Little Creek Mountain southeast of Hurricane. This 8-km-long group consists of two closely spaced alignments with an average trend of N. 40° W. Some centers are low, gently sloping lava mounds flanked or partially flanked by aprons of scoria and volcanic bombs. The cones and mounds in this alignment exhibit similar degrees of erosional modification, suggesting that they erupted quasi-simultaneously, probably from a single magmatic source. Most of the vents in the group represent eruptions of very small volume. The two largest-volume vents are marked by conspicuous cones whose volumes may have special structural significance. One is located along Gould Creek at the northwest end of the alignment at an elevation 250 m lower than those on Little Creek Mountain. This cone is built on several lava flows that erupted from the vent and flowed as much as 7 km

westward toward the Hurricane Cliffs, exceeding in total volume the eruptions from the remainder of the aligned vents. These large-volume eruptions from the topographically lowest point in the alignment suggest integrated magmatic geopressure over the length of the alignment. The second-largest cone is located at the intersection of the alignment with a northeast-striking steep fault, suggesting enhancement of magma ascent by the damage zone of the cross fault. Steep-walled canyons cut into Little Creek Mountain as well as its plateau-like edges provide excellent exposures of pre-basalt rocks transverse to the aligned vents. The eruptions were controlled by a vertical joint striking N. 40° W. along which there is no shear displacement. A 3-m-wide basalt dike fills the joint at one locality but elsewhere there is no indication of significant dilational strain resulting in the formation of dikes or veins. Either the walls of the joint closed following eruptions or many small tube-like channelways are buried beneath the vents. We conclude that these eruptions were not accompanied by synvolcanic faulting or hazardous levels of seismicity.

On the Skutumpah Terrace east of Glendale and Orderville, three basaltic cones form conspicuous landmarks: Buck Knoll, Black Knoll, and Bald Knoll (plate 1). None are on mapped faults, though Bald Knoll, the eastern one, is within 0.5 km of the projected trace of a mapped fault (Doelling and Davis, 1989). Black Knoll is near the northeast termination of a major steep joint that is part of an extensive, young, northeast-striking joint system rendered conspicuous by spectacular erosional etching of Navajo Sandstone (figure 17). As with the joint at Little Creek Mountain, excellent exposures show that there is no shear displacement or significant permanent dilational strain at the joint. We assume that eruptions at Black Knoll were not accompanied by faulting or significant seismic-

Figure 17. Photo looking southwest across canyon in headwater area of Kanab Creek about 4 km southwest of Black Knoll showing joint-controlled erosional notches in Navajo Sandstone. The joint that controls the notch on the right can be traced 10 km to the southwest from here. To the northeast, it projects into the volcanic cone at Black Knoll.

ity. Buck Knoll is probably also on a joint as indicated by a lack of fault displacement and the presence of aligned vegetation north of the cone. Thus, even some individual cones appear to be joint controlled.

Aligned vents in the Cedar City quadrangle are generally not on mapped faults. For those aligned along northeast trends, parallel-striking faults of known or inferred Quaternary age cut adjacent rocks or rocks along the projection of the aligned vents. The northwest-aligned vents at Little Creek Mountain have no parallel-striking Quaternary faults nearby, but the north-northwest-aligned vents on the Uinkaret Plateau in the adjacent part of Arizona are paralleled by faults, some of which have been active during Quaternary time (Koons, 1945). As with the vents on the Markagunt Plateau, those that form alignments on the Uinkaret are not located along mapped faults. Also, aligned and indistinctly aligned cinder cones in the region north of St. George generally are not located on mapped faults. This general lack of association of Quaternary volcanism with faults, together with direct evidence for joint-controlled volcanism, leads us to speculate that Quaternary volcanism in the Cedar City quadrangle was generally not accompanied by seismogenic faulting at hazardous levels.

SUMMARY

The principal through-going Quaternary structures in the Cedar City quadrangle are the Sevier fault and the Hurricane fault. Both are generally north-northeast-striking normal faults and are spacially and temporally associated with local smaller scale faulting and folding on the hanging wall (downthrown) blocks. The Hurricane fault has greater long-term slip rates and activity than the Sevier fault, although both have been active in the late Pleistocene. Probable late Pleistocene scarps on alluvium and pediment/colluvial deposits are on the Hurricane fault, but no scarps on alluvium are on the Sevier fault. Late Pleistocene displacement on the Sevier fault is inferred solely from indirect geomorphic evidence and from presumed age protraction of many small displacements to produce large offsets in basalt of middle Pleistocene age.

Other major structures in the quadrangle are less continuous, shorter, and generally older. The Gunlock fault has been relatively inactive, with the last movement occurring during early to middle Pleistocene time. This fault represents the short northern part of the larger Grand Wash fault system in Arizona and is the only part within Utah of this long fault system which shows evidence for Quaternary movement. The Washington fault likewise extends northward into Utah from Arizona, but in Utah it is generally a middle to late Pleistocene structure with local evidence for late Pleistocene displacement near Warner Ridge. The Antelope Range fault which bounds the eastern edge of the Escalante Desert has scarps on alluvium. Even though recurrent movement on this generally active mountain front has occurred in the past, this fault is considered to have been quiescent during Holocene time because of the advanced state of degradation of the scarps.

Many other Quaternary structures occur in the quadrangle. Most trend north-northeast parallel to major structures, but a few isolated features trend east-west. The youngest surface-faulting events are in the northern part of the quadrangle: (1)

north of Cedar City in the Enoch-Parowan Valley area, (2) in the northern Escalante Desert, and (3) in the Sevier Valley north of Panguitch. Scarp profiles and exposures in the Enoch-Parowan Valley area indicate a latest Pleistocene (possible Holocene) age for some scarps. To the west in the Escalante Desert near Zane and Avon, and north of Panquitch in the Sanford Creek area, faults with possible late Pleistocene displacement are found.

POTENTIAL FOR SURFACE FAULTING

Our assessment of the potential for surface rupture on faults in the Cedar City quadrangle is a qualitative one based solely on geologic information regarding Quaternary faulting. No attempt is made to quantify overall probabilities for the area based either on an analysis of historical seismicity or on a relationship between long-term geologic slip rates and seismic moment rates. Such analyses for southwestern Utah, done for an area about twice as large as the Cedar City quadrangle which includes the more active Richfield area, indicate an estimated recurrence for an M_L 7.0-7.5 earthquake somewhere in the region of 200-500 years based on historical seismicity (Earth Science Associates, 1982) and 200-600 years based on geologically determined moment rates (Doser and Smith, 1982). These analyses, however, do not address the specific likelihood of a surface-faulting event at a particular place on a fault.

Although no coseismic surface faulting has been reported during historical time in the Cedar City quadrangle, many faults are present on which surface faulting has occurred repeatedly during late Quaternary time. Detailed investigations have not been undertaken to evaluate the time of last movement, recurrence intervals, displacement per event, and segmentation of these faults; studies of this type are necessary to fully evaluate the likelihood of surface faulting. Our preliminary assessment based on these reconnaissance investigations is that the likelihood of surface faulting on any particular fault in the quadrangle is relatively low. Although long-term slip rates on some faults are high and within an order of magnitude of the more active Wasatch fault, these faults have had little, if any, displacement during Holocene time. Little data are available regarding the amount of offset during individual surface-faulting events on major faults. Trenching studies on the Washington fault indicate that offsets exceeding 1 m may have occurred during single events. On minor faults such as those near Panquitch at Sanford Creek, the offset per event was less than 1 m.

Variations in trend, prominent cross-structures, and structural-inheritance features along major faults indicate that segments with independent faulting histories may be defined with further work. We found no obvious differences in apparent time of last movement on individual faults, but this is probably because of the long elapsed time since the last event. Unfortunately, the major faults have few sites where detailed trenching studies are likely to yield reliable data that would define recurrence intervals and the timing of the last event.

No faults in this quadrangle show evidence for repeated surface rupture during Holocene time, similar to the Wasatch fault in northern Utah. The faults with evidence for most recent movement are generally short, discontinuous, and do not bound major range fronts. The Hurricane and Paragonah faults probably pose the greatest threat for generating a large-magnitude surface-faulting earthquake in a populated area.

Their relatively high long-term slip rates contrast with the general lack of evidence for recurrent Holocene movement. The slip rates indicate that for thousand-year-plus exposure times the probability of a large-magnitude surface-faulting earthquake is relatively high. Although it is possible that a surface-faulting earthquake could occur on any fault at any time, the apparent long recurrence intervals and lack of activity during the Holocene indicate that the probability of a large surface-faulting earthquake at a particular point on any of these faults within the time frame generally considered for most engineering works (50-100 years) is low, particularly relative to the Wasatch fault and others in northern Utah.

ACKNOWLEDGEMENTS

We thank Peter Rowley, Edward Sable, Kenneth Hamblin, and Hellmut Doelling for providing us with pre-publication copies of maps of the Cedar Valley, High Plateau, St. George, and Kane County areas, respectively. We thank Kathy Haller, Suzanne Hecker, and Meridee Jones-Cecil for very helpful reviews of the manuscript. Kenneth Hulet volunteered assistance in measuring scarp profiles in the Pintura area and Mary Anderson did so in several other areas. Mary Anderson volunteered help in sampling some carbon-bearing sites. Monte Neilson provided occasional logistic support such as loans of tools, chain saw, and motorcycle and shuttle service. Bill D. Black drafted the figures. Anderson is grateful for the hospitality extended him by the Randy Wilkinson and Merlin Baker families of St. George and the Monte Neilson and Belle Hulet families of Cedar City.

REFERENCES

Anderson, J.J., and Rowley, P.D., 1975, Cenozoic stratigraphy of southwestern High Plateaus of Utah, *in* Anderson, J.J., and others, editors, Cenozoic Geology of Southwestern High Plateaus of Utah, Geological Society of America Special Paper 160, p. 1-51.

Anderson, J.J., and Rowley, P.D., 1987, Geologic map of the Panguitch NW Quadrangle, Iron and Garfield Counties, Utah: Utah Geological and Mineral Survey Map 103, scale 1: 24,000.

Anderson, J.J., Rowley, P.D., Fleck, R.J., and Nairn, A.E.M., 1975, Cenozoic geology of southwestern high plateaus of Utah: Geological Society of America Special Paper 160, 88 p.

Anderson, J.J., Iivari, T.A., and Rowley, P.D., 1987, Geologic map of the Little Creek Peak quadrangle, Garfield and Iron Counties, Utah: Utah Geological and Mineral Survey Map 104, scale 1: 24,000.

Anderson, L.W., and Miller, D.G., 1979, Quaternary fault map of Utah: Long Beach, California, Fugro, Inc., 32 p., scale 1: 500,000.

Anderson, R.E., 1978, Quaternary tectonics along the Intermountain seismic belt south of Provo, Utah: Brigham Young University Geology Studies, v. 25, pl. 1, p. 1-10.

Anderson, R.E., 1980, The status of seismotectonic studies of southwestern Utah: U.S. Geological Survey Open-File Report 80-801, p. 519-547.

Anderson, R.E., and Barnhard, T.P., 1987, Neotectonic framework of the central Sevier Valley area, Utah, and its relationship to seismicity: U.S. Geological Survey Open-File Report 87-585, p. F1-F134.

Anderson, R.E., and Bucknam, R.C., 1979a, Map of fault scarps on unconsolidated sediments, Richfield 1° x 2° Quadrangle, Utah: U.S. Geological Survey Open-File Report 79-1236, 15 p.

Anderson, R.E., and Bucknam, R.C., 1979b, Two areas of Holocene deformation in southwestern Utah: Tectonophysics, v. 52, p. 417-430.

Anderson, R.E., Bucknam, R.C., and Hamblin, Kenneth, 1978, Road log to the Quaternary tectonics of the Intermountain Seismic Belt between Provo and Cedar City, Utah: Geological Society of America, Rocky Mountain Section Field Trip No. 8, Provo, Utah, April 28-29, 1987, 50 p.

Anderson, R.E., and Mehnert, H., 1979, Reinterpretation of the history of the Hurricane fault in Utah: Rocky Mountain Association of Geologists—Utah Geological Association, 1979 Basin and Range Symposium, p. 145-173.

Arabasz, W.J., and Julander, D.R., 1986, Geometry of seismically active faults and crustal deformation within the Basin and Range-Colorado Plateau transition in Utah, *in* Mayer, Larry, editor, Extensional Tectonics of the Southwestern United States: A Perspective on Processes and Kinematics: Geological Society of America Special Paper 208, p. 43-74.

Arabasz, W.J., and Smith R.B., 1979, The November 1971 earthquake swarm near Cedar City, Utah, *in* Arabasz, W.J., Smith, R.B., and Richins, W.D., editors, Earthquake Studies in Utah, 1850-1978: Salt Lake City, Utah, University of Utah Seismograph Stations, Department of Geology and Geophysics, p. 423-432.

Arabasz, W.J., Smith, R.B., and Richins, W.D., editors, 1979, Earthquake studies in Utah, 1850-1978: Salt Lake City, Utah, University of Utah Seismograph Stations, Department of Geology and Geophysics, 552 p.

Armstrong, R.L., 1968, Sevier orogenic belt in Nevada and Utah: Geological Society of America Bulletin, v. 79, p. 429-458.

Averitt, Paul, 1962, Geology and coal resources of the Cedar Mountain quadrangle, Iron County, Utah: U.S. Geological Survey Professional Paper 389, 72 p.

——1967, Geologic map of the Kanarraville quadrangle, Iron County, Utah: U.S. Geological Survey Geologic Quadrangle Map GQ-694, 1: 24,000.

Beck, A.C., 1968, Gravity faulting as a mechanism of topographic adjustment: New Zealand Geological Survey, Department of Scientific and Industrial Research, Christchurch, New Zealand.

Best, M.G., and Brimhall, W.H., 1974, Late Cenozoic alkalic basaltic magmas in the western Colorado plateaus and the Basin and Range transition zone, U.S.A., and their bearing on mantle dynamics: Geologic Society of America Bulletin, v. 85, p. 1677-1690.

Best, M.G., McKee, E.H., and Damon, P.E., 1980, Space-time-composition patterns of late Cenozoic mafic volcanism, southwestern Utah and adjoining areas: American Journal of Science, v. 280, p. 1035-1050.

Bucknam, R.C., and Anderson, R.E., 1979, Estimation of fault-scarp ages from scarp-height-slope-angle relationship: Geology, v. 7, p. 11-14.

Burchfiel, B.C., and Hickcox, C.W., 1972, Structural development of central Utah, *in* Baer, J.L., and Callaghan, E. editors, Plateau-Basin and Range Transition Zone, Central Utah, 1972: Salt Lake City, Utah Geological Association Publication 2, p. 55-56.

Carpenter, C.H., Robinson, G.B., and Bjorklund, L.J., 1967, Ground-water conditions and geologic reconnaissance of the upper Sevier River Basin, Utah: U.S. Geological Survey Water-Supply Paper 1836, 91 p.

Cashion, W.B., 1961, Geology and fuels resources of the Orderville-Glendale area, Kane County, Utah: U.S. Geological Survey Coal Investigations Map C-49, scale 1: 62,500.

Cashion, W.B., 1967, Geologic map of the south flank of the Markagunt Plateau northwest Kane County, Utah: U.S. Geological Survey Miscellaneous Geologic Investigations Map I-494.

Christenson, G.E., and Deen, R.D., 1983, Engineering geology of the St. George area, Washington County, Utah: Utah Geological and Mineral Survey Special Studies 58, 32 p.

Cook, E.F., 1957, Geology of the Pine Valley Mountains, Utah: Utah Geological and Mineralogical Survey Bulletin 58, 114 p.

Cook, E.F., 1960, Geologic atlas of Utah—Washington County: Utah Geological and Mineralogical Survey Bulletin 70, 124 p.

Cook, K.L., and Hardyman, E., 1967, Regional gravity survey of the Hurricane fault area and Iron Springs district, Utah: Geological Society of America Bulletin, v. 78, p. 1063-1076.

Davis, G.H., 1980, Structural characteristics of metamorphic core complexes, southern Arizona, *in* Crittenden, M.D., Coney, P.J., and Davis, G.H., editors, Cordilleran Metamorphic Core Complexes: Geological Society of America Memoir 153, p. 35-77.

Doelling, H.H., 1975, Geology and mineral resources of Garfield County, Utah: Utah Geological and Mineral Survey Bulletin 107, 175 p.

Doelling, H.H., and Davis, F.D., 1989, The geology of Kane County, Utah: Utah Geological and Mineral Survey Bulletin 124, 194 p.

Doser, D.I., and Smith, R.B., 1982, Seismic moment rates in the Utah region: Seismological Society of America Bulletin, v. 72, p. 525-551.

Earth Science Associates, 1982, Seismic safety investigation of eight SCS dams in southwestern Utah: Unpublished consultant's report for the U.S. Soil Conservation Service, Portland, Oregon, 2 vols.

Ebree, G.F., Lateral and vertical variations in a Quaternary basalt flow; Petrography and chemistry of the Gunlock flow, southwestern Utah: Brigham Young University Geology Studies, v. 17, pt. 1, p. 67-115.

Ertec Western, Inc., 1981, Faults and lineaments in the MX siting region, Nevada and Utah: Unpublished consulting report for U.S. Air Force, 77 p.

Fitton, Godfrey, and James, Dodie, 1986, Regional compositional variation among late Cenozoic basalts in the southwestern USA: Geological Society of America Abstracts with Programs, v. 18, no. 6, p. 602.

Fleck, R.J., Anderson, J.J., and Rowley, P.D., 1975, Chronology of mid-Tertiary volcanism in High Plateaus region of Utah, *in* Cenozoic Geology of Southwestern High Plateaus of Utah: Geological Society of America Special Paper 160, p. 53-62.

Gile, L.H., Peterson, F.F., and Grossman, R.B., 1966, Morphological and genetic sequences of carbonate accumulation in desert soils: Soils Science, v. 101, p. 347-360.

Grant, S.K., and Proctor, P.D., 1988, Geologic map of the Antelope Peak Quadrangle, Iron County, Utah: Utah Geological and Mineral Survey Open-File Report 130, scale 1:24,000.

Hamblin, W.K., 1965, Origin of "reverse drag" on the downthrown side of normal faults: Geological Society of America Bulletin, v. 76, p. 1145-1164.

——1970a, Structure of the western Grand Canyon region, *in* Hamblin, W.K., and Best, M.G., editors. The Western Grand Canyon District: Utah Geological Society Guidebook to the Geology of Utah, no. 23, p. 3-19.

——1970b, Late cenozoic basalt flows of the western Grand Canyon, *in* Hamblin, W.K., and Best, M.G., editors, The Western Grand Canyon District: Utah Geological Society Guidebook to the Geology of Utah, no. 23, p. 21-37.

——1976, Patterns of displacement along the Wasatch fault: Geology v. 4, p. 619-622.

Hamblin, W.K., Damon, P.E., and Bull, W.B., 1981, Estimates of vertical crustal strain rates along the western margins of the Colorado Plateau: Geology, v. 9, p. 293-298.

Hanks, T.C., Bucknam, R.C., Lajoie, K.R., and Wallace, R.E., 1984, Modification of wave-cut and faulting-controlled landforms: Journal of Geophysical Research, v. 89, p. 5771-5790.

Hintze, L.F., compiler, 1980, Geologic map of Utah: Utah Geological and Mineral Survey Map, scale 1:500,000.

——1986, Stratigraphy and structure of the Beaver Dam Mountains, southwestern Utah: Utah Geological Association Publication 15, p. 1-36.

Holmes, R.D., 1979, Thermoluminescence dating of Quaternary basalts: continental basalts from the eastern margin of the Basin and Range province, Utah and northern Arizona: Brigham Young University, Geology Studies, v. 26, pt. 2, p. 51-65.

Koons, E.D., 1945, Geology of the Uinkaret plateau, northern Arizona: Geological Society of America Bulletin, v. 56, p. 151-180.

Kurie, A.E., 1966, Recurrent structural disturbance of Colorado Plateau margin near Zion National Park, Utah: Geological Society of America Bulletin, v. 77, p. 867-872.

Luedke, R.G., and Smith, R.L., 1978, Map showing distribution, composition, and age of late Cenozoic volcanic centers in Colorado, Utah, and southwestern Wyoming: U.S. Geological Survey Miscellaneous Geologic Investigations Map I-1091-B, scale 1:1,000,000.

Machette, M.N., 1985a, Late Cenozoic geology of the Beaver basin, southwestern Utah: Brigham Young University, Geology Studies, v. 32, pt. 1, p. 19-37.

Machette, M.N., 1985b, Calcic soils of the southwestern United States, *in* Weide, D.L., editor, Soils and Quaternary Geology of the Southwestern United States: Geological Society of America Special Paper 203, p. 1-21.

Machette, M.N., Personius, S.F., Menges, C.M., and Pearthree, P.A., 1986, Map showing Quaternary and Pliocene faults in the Silver City 1° X 2° quadrangle and the Douglas 1° X 2° quadrangle, southeastern Arizona and southwestern New Mexico: U.S. Geological Survey Miscellaneous Field Studies Map MF-1465-C.

Mackin, J.H., and Rowley, P.D., 1975, Geologic map of the Avon SE Quadrangle, Iron County, Utah: U.S. Geological Survey Geologic Quadrangle Map GQ-1294, scale 1:24,000.

——1976, Geologic Map of the Three Peaks Quadrangle, Iron County, Utah: U.S. Geological Survey Geologic Quadrangle Map GQ-1297, scale 1:24,000.

Menges, C.M., and Pearthree, P.A., 1983, Map of neotectonic (latest Pliocene-Quaternary) deformation in Arizona: Arizona Bureau of Geology and Mineral Technology Open-File Report 83-22, 15 p.

Naeser, C.W., Bryant, B.R., Crittenden, M.D., and Sorenson, M.L., 1980, Fission-track dating in the Wasatch Mountains, Utah—An uplift study, *in* Andriese, P.D., compiler, Proceedings of Conference X Earthquake Hazards along the Wasatch and Sierra Nevada Frontal Fault Zones: U.S. Geological Survey, Open-File Report 80-801, p. 634-646.

Nash, D.B., 1980, Morphologic dating of degraded normal fault scarps: Journal of Geology, v. 88, p. 353-360.

Nielson, R.L., 1983, Pleistocene and Holocene evolution of Little Salt Lake playa (Lake Parowan), southwest Utah: Geological Society of America Abstracts with Programs, v. 15, no. 5, p. 300.

Peterson, S.M., 1983, The tectonics of the Washington fault zone, northern Mojave County, Arizona: Brigham Young University Geology Studies, v. 30, pl. 1, p. 83-94.

Radbruch-Hall, D.H., Varnes, D.J., and Colton, R.B., 1977, Gravitational spreading of steep-sided ridges ("sackung") in Colorado:

Journal of Geophysical Research, U.S. Geological Survey, v. 5, no. 3, p. 359-363.

Richins W.D., Zandt, G., and Arabasz, W.J., 1981, Swarm seismicity along the Hurricane fault zone during 1980-1981: A typical example for SW Utah: EOS, Transactions of the American Geophysical Union (abs.), v. 62, no. 45, p. 966.

Rowley, P.D., 1975, Geologic map of the Enoch NE quadrangle, Iron County, Utah: U.S. Geological Survey Geologic Quadrangle map GQ-1301, scale 1:24,000.

Rowley, P.D., 1976, Geologic map of the Enoch NW quadrangle, Iron County, Utah: U.S. Geological Survey Geologic Quadrangle Map GQ-1302, scale 1:24,000.

Rowley, P.D., Anderson, J.J., Williams, P.L., and Fleck, R.J., 1978, Age of structural differentiation between the Colorado Plateaus and Basin and Range provinces in southwestern Utah: Geology, v. 6, p. 51-55.

Rowley, P.D., Hereford, Richard, and Williams, V.S., 1987, Geologic map of the Adams Head-Johns Valley area, southern Sevier Plateau, Garfield County, Utah: U.S. Geological Survey Miscellaneous Investigations Series Map I-1798, scale 1:50,000.

Rowley, P.D., Steven, T.A., and Kaplan, A.M., 1981, Geologic map of the Monroe NE quadrangle, Sevier County, Utah: U.S. Geological Survey Miscellaneous Field Studies Map MF-1330, scale 1:24,000.

Rowley, P.D., and Threet, R.L., 1976, Geologic map of the Enoch quadrangle, Iron County, Utah: U.S. Geological Survey Geologic Quadrangle Map GQ-1296, scale 1:24,000.

Sawyer, R.F., and Cook, K.L., 1977, Gravity and ground magnetic surveys of the Thermo Hot Springs KGRA region, Beaver County, Utah: University of Utah, Department of Geology and Geophysics, Tech-77-6, 142 p.

Scarborough, R.B., Menges, C.M., and Pearthree, P.A., 1986, Late Pliocene-Quaternary (post-4 m.y.) faults, folds, and volcanic rocks in Arizona: Arizona Bureau of Geology and Mineral Technology Map 22, scale 1:1,000,000.

Segall, P., and Pollard, D.D., 1980, Mechanics of discontinuous faults: Journal of Geophysical Research, v. 85, p. 4337-4350.

Shubat, M.A., and Siders, M.A., 1988, Geologic map of the Silver Peak quadrangle, Iron County, Utah: Utah Geological and Mineral Survey Map 108, scale 1:24,000.

Sibson, R.H., 1985, Stopping of earthquake ruptures at dilational fault jogs: Nature, no. 316, p. 248-251.

Siders, M.A., Rowley, P.D., Shubat, M.A., Christenson, G.E., and Galyardt, G.L., 1989, Geologic map of the Newcastle quadrangle, Iron County, Utah: U.S. Geological Survey Open-File Report 89-449, scale 1:24,000.

Smith, R.B., 1979, Fundamentals of earthquakes seismology, *in* Arabasz, W.D., Smith, R.B., and Richins, W.D., editors, Earthquake Studies in Utah, 1850-1978: Salt Lake City, Utah, University of Utah Seismograph Stations, Department of Geology and Geophysics, p. 15-32.

Smith, R.B., and Sbar, M.L., 1974, Contemporary tectonics and seismicity of the western United States with emphasis on the Intermountain seismic belt: Geological Society of America Bulletin, v. 85, p. 1205-1218.

Smith, R.L., and Luedke, R.G., 1984, Potentially active volcanic lineaments and loci in western conterminous United States, *in* Explosive Volcanism, Inception, Evolution, and Hazards: National Academy Press, 176 p.

Stewart, J.H., Moore, W.J., and Zietz, Isidore, 1977, East-west patterns of Cenozoic igneous rocks, aeromagnetic anomalies, and mineral deposits, Nevada and Utah: Geological Society of America Bulletin, v. 88, p. 67-77.

Swan, F.H., III, Schwartz, D.P., and Cluff, L.S., 1980, Recurrence of moderate to large magnitude earthquakes produced by surface faulting on the Wasatch fault, Utah: Bulletin of the Seismological Society of America, v. 70, no. 5, p. 1431-1462.

Threet, R.L., 1963, Geology of the Parowan Gap area, Iron County, Utah, *in* Geology of Southwestern Utah: Intermountain Association of Petroleum Geologists, 12th Annual Field Conference 1963, Utah Geological and Mineralogical Survey, p. 136-145.

Wagner, J.J., 1984, Geology of the Haycock Mountain 7.5 minute quadrangle, western Garfield County, Utah: Kent, Ohio, Kent State University, M.S. thesis, 68 p.

Wallace, R.E., 1977, Profiles and ages of young fault scarps, north-central Nevada: Geological Society of America Bulletin, v. 88, p. 1267-1281.

Wheeler, R.L., and Krystinik, K.B., 1988, Segmentation of the Wasatch fault zone, Utah-summaries, analysis, and interpretations of geological and geophysical data: U.S. Geological Survey Bulletin 1827, 47 p.

THE UTAH GEOLOGICAL AND MINERAL SURVEY is one of eight divisions in the Utah Department of Natural Resources. The UGMS inventories the geologic resources of Utah (including metallic, nonmetallic, energy, and ground-water sources); identifies the state's geologic and topographic hazards (including seismic, landslide, mudflow, lake level fluctuations, rockfalls, adverse soil conditions, high ground water); maps geology and studies the rock formations and their structural habitat; and provides information to decisionmakers at local, state, and federal levels.

THE UGMS is organized into five programs. Administration provides support to the programs. The Economic Geology Program undertakes studies to map mining districts, to monitor the brines of the Great Salt Lake, to identify coal, geothermal, uranium, petroleum and industrial minerals resources, and to develop computerized resource data bases. The Applied Geology Program responds to requests from local and state governmental entities for site investigations of critical facilities, documents, responds to and seeks to understand geologic hazards, and compiles geologic hazards information. The Geologic Mapping Program maps the bedrock and surficial geology of the state at a regional scale by county and at a more detailed scale by quadrangle.

THE INFORMATION PROGRAM distributes publications, answers inquiries from the public, and manages the UGMS Library. The UGMS Library is open to the public and contains many reference works on Utah geology and many unpublished documents about Utah geology by UGMS staff and others. The UGMS has begun several computer data bases with information on mineral and energy resources, geologic hazards, and bibliographic references. Most files are not available by direct access but can be obtained through the library.

THE UGMS PUBLISHES the results of its investigations in the form of maps, reports, and compilations of data that are accessible to the public. For future information on UGMS publications, contact the UGMS sales office, 606 Black Hawk Way, Salt Lake City, Utah 84108-1280.
